2197

D0426403

Dr John Medina is a molecular biologist on the faculty of the Department of Bioengineering at the University of Washington School of Medicine. He is also a consultant and regular columnist for an American psychiatric organization on the genetics of human behavior. In the course of his research career, which has included the isolation and characterization of genes involved in cardiovascular development, Dr Medina became concerned with the public communication of biological science to a lay audience. He is as much at ease writing to a non-specialist audience as he is authoring a paper for a research journal. His books include *The Outer Limits of Life* and *Uncovering the Mystery of Aids*, published by Thomas Nelson.

Anyone who has watched a wrinkle slowly gouge their face like a strip mine, or has been disturbed by a loss of memory, has uncomfortably confronted the human aging process. The inexorable march of time on our bodies begs an important question: why do we have to grow old? Written in everyday language, *The Clock of Ages* takes us on a tour of the aging human body – all from a research scientist's point of view. From the deliberate creation of organisms that live three times their natural span to the isolation of human genes that may allow us to do the same, *The Clock of Ages* also examines the latest discoveries in geriatric genetics. Sprinkled throughout the pages are descriptions of the aging of many historical figures, such as Florence Nightingale, Jane Austen, Bonaparte and Casanova. These stories underscore the common bond that unites us all: they aged, even as we do. *The Clock of Ages* tells us why.

Related titles

The Thread of Life
SUSAN ALDRIDGE
Remarkable Discoveries!
FRANK ASHALL
Evolving the Mind
A. G. CAIRNS-SMITH
Extraterrestrial Intelligence
JEAN HEIDMANN
Hard to Swallow: a brief history of food
RICHARD LACEY
An Inventor in the Garden of Eden
ERIC LAITHWAITE
The Outer Reaches of Life
JOHN POSTGATE
Prometheus Bound:
science in dynamic steady state
JOHN ZIMAN

JOHN J. MEDINA

The Clock
of Ages

Why we age – how we age – winding back the clock

CAMBRIDGE
UNIVERSITY PRESS

Published by the Press Syndicate of the University of Cambridge
The Pitt Building, Trumpington Street, Cambridge CB2 1RP
40 West 20th Street, New York, NY 10011-4211, USA
10 Stamford Road, Oakleigh, Melbourne 3166, Australia

© Cambridge University Press 1996

First published 1996

Printed in Great Britain by Biddles Ltd, Guildford & King's Lynn

A catalogue record for this book is available from the British Library

Library of Congress cataloguing in publication data
Medina, John J.
The clock of ages / John J. Medina.
p.   cm.
Includes index.
ISBN 0 521 46244 4 (hc)
1. Aging.   2. Aging – Genetic aspects.   3. Aging – Molecular
aspects.   I. Title.
QP86.M52   1996
612.6'7 – dc20   95–40712 CIP

ISBN 0 521 46244 4 hardback

Contents

Preface

It was time to say my last words to my mother.

She was dying. Not much had changed since I had last seen her. She still had a full head of hair, making her look much younger than her 64 years. Her voice betrayed some of her tenure, though. It was almost half an octave higher than the one I had heard as a little boy, the product of a natural stiffening of the vocal cords. The lines on her face spoke of her years too, already sculpted by the finger of time, greatly deepened by decades of loving laughter. These marks always concerned her, though she had once read that wrinkles were a natural, unstoppable part of growing older. She often looked in the bathroom mirror – even as a young mother – to examine their progression. 'The Clock of Ages', I would sing to her at the top of my lungs, making a pun from a hymn she loved to hear at church. She paused. 'But not cleft for me, unfortunately,' she sighed, tilting her head for the hundredth time, still looking in the mirror.

When I came to see her, she was lying in her bed. It was a darkened room, lit only by the soundless snow from an unwatched TV. I turned off the attached video tape machine. She had fallen asleep watching an old black and white movie from Hollywood.

'They liked young faces, you know,' she told me once in high school. In the days when the wrinkles did not exist, Mom had been a promising young actress at the University of Michigan. And much more. She was on the diving team, in the choir, inducted into the honors society, one of those people that kind of make you sick at graduation because they get all these awards. And impressive all the more because she was so nice, always laughing, always giving away a smile you'd remember all day, as effervescent as a can of soda pop.

Where she truly shone, however, was in the talent of her chosen major, the dramatic arts. Her gift was of sufficient quality that she began starring in stage productions with the likes of Fernando Llamas and Ricardo Montalban. She started a correspondence with Basil Rathbone and eventually with Jane Wyman and Ronald Reagan. In going through her things I found a number of letters and a ream of publicity photos I didn't know existed. She was so *young*.

'They told me that I should go to Hollywood before I got too old.' Her eyes twinkled as she related an earlier time for me. Now I was going off to college and she wanted to give me some advice. 'That's because there were almost no parts for women over the age of 30. To be honest, John, there are still no parts for older women!' Taking the advice of her coach and a few famous friends, she went to Hollywood before The Clock of Ages took its toll.

Mom was immediately successful, of course. Her connections on the stage got her a screen test for a part in an upcoming movie, and she outdid all her competition. 'The day I found I was going to get the part was the happiest day of my life.' Mom said, 'And then also the saddest.' The young starlet was called into one of the executive's offices to talk about negotiating a contract. He wasn't well known to the outside world, but he was a power broker in the feudal star system of the early 1950s. And when he began to talk about the problems he and his wife were having and how beautifully youthful Mom looked and how this contract would be so easy to offer if Mom would be nice to him and how smooth life would be for future projects if she remained friendly to him and how awful it would be if she said no . . .

'I knew inside what he was doing, John. I also knew that if I refused his offer, that would be the end of my career,' she told me. 'I was shocked, of course. This kind of thing did not happen in Michigan. And I was raised in a different time. So I thought about it there in his office. In the end, it was more important that I be able to look at my image in the mirror every morning than gaze at my image on the silver screen every night.' With a sinking feeling that only comes when you are watching life-long hopes and dreams die at your own hand, she refused his offer.

The executive told her that she was a fool, that for every young girl in her position there were ten waiting in line to take her place. Mom said that this was fine, and told him to give the other ten a phone call, because she was walking out the door. 'Now I have these photographs, and a couple of letters,' she said as she gathered them into an envelope. And then wearing a slight smile, she gave some of them to me before I went off to college. And then she left the room.

But that was a long time ago. And also several worlds away. Now I am an adult and I am watching this would-be movie star and life-long friend sputter against a Clock she declared war on many years ago. The doctors had told us she was dying. But we knew long before. She didn't talk very much at the end, except to whisper for an occasional glass of water. And when we gave it to her she said thank-you and smiled.

Try as I might, I wasn't there to recollect old memories with her. I was there to say my last words and then leave, because I am a research professor at a medical school and I will have to drive all night to get back to the laboratory. I draw near to the side of her bed and pray in the dark she cannot see me crying. I can see her, however, her gaunt face outlined in the flicker of the unwatched TV. I manage to stammer 'Tell Jesus I said hi, okay?' and I touch her hand and I hear her waken. 'Okay,' she whispers hoarsely and then she pauses. In a slightly louder voice, she asks:

'Will you please turn off the television?'

Mom slipped into unconsciousness shortly after that and died seven days later.

The purpose of this book

No laboratory experience can ever prepare you for the death of a loved one. You can anticipate and plan and rehearse your feelings for an Oscar-winning performance, and then when death comes, you mostly just stammer. The impending sense of loss is extreme, the helplessness in the face of biological processes you can't control is frustrating and terrifying.

The death of relatives and friends is powerful for another reason too. Lying in the back of your mind, like a sleeping dog, is the raw fact of your own mortality. If someone close to you can age and die, then you can too. As the years pass, certain immutable changes occur in our bodies, which do nothing but remind us of this strange terminal outcome. Such a sense of timed fragility is one of the most profound feelings I have ever experienced. It is the persistent ticking of The Clock of Ages.

The book you have in your hands describes a personal tour. Not through a strange country, or even a foreign concept. It is a tour through a biological time piece, this amazing Clock of Ages, which is the aging process in human beings. The tour is informational. It will be filled with various stops around the world and, eventually, around our bodies. We will pause to consider questions at specific times to enlighten us on the aging process, questions like 'Why do different animals have different life spans?', 'Why does my hair turn gray?' and 'Am I really losing my memory?' Because I am a scientist as well as the tour guide, we will approach the answer to these questions from a reductionist's point of view; that is, we will look closely at the unimaginable complex world of our tissues and cells and even genes. These are the objects, of course, that make up the screws and gears and springs of the Clock. We will certainly observe the machine in operation, and we will even speculate on work that may have found a way to extend its operation. Or in a few cases, even wind it back.

The first stop on our tour is a historical one. We will attempt to place aging and death into an evolutionary context. Why do organisms age at all? Is our ability to reproduce tied to our life span? Are there creatures that never die? To answer these questions, we will first have to outline some operating definitions of death in the general and human death in the specific. As we'll see, that may not be very easy to do.

The second stop on our tour is a description of the various components of the Clock. We will describe how the divers tissues and organs of the human body change over time. Why does my skin wrinkle as I get older? What happens to my thinking processes as I age? Why must I place books farther from my eyes in order to see the text? By individually examining these disparate systems, we will discover the effects of the years on our capabilities – and to what we have to look forward as we age.

The last stop on our tour is a description of how the components work together to make the Clock tick. Instead of examining whole tissue and organs, we will observe individual cells and the lilliputian genes they contain. Are there genes that deliberately cause cells to die? Are there genes that can extend their life span? What can I do to arrest or even turn back the effects of aging on my body? To answer these questions, we will discuss how various genes work inside cells and creatures. We will discuss some exciting new frontiers in the effort to rewind The Clock of Ages.

It is fully realized by this reductionist scientist that aging and dying have more components than just test tubes and petri dishes. These processes are a shared experience that in many ways fertilizes our religions and causes entire societies to behave in certain ways. The reason, of course, is that The Clock of Ages possesses a ticking heard by everyone. Included in this tour package, then, is a description of how other historically familiar human beings have experienced aging and death. We will learn about the attitudes of writers like Jane Austen and painters like

Francisco Goya, nurses like Florence Nightingale and generals like Napoleon Bonaparte, lovers like Casanova and criminals like Billy the Kid. They have in common the fact of facing this Clock head on, even as we do today. They have in addition the fact that they experienced its final ticking, even as we have not.

A quick word about expertise

Even though this text is about science and biology, please don't let that fact be frightening. The tour guide is well aware that even if we share a similar biological fate, we do not similarly share biological backgrounds. You can have flunked most of your grade school biology and still understand everything written in these pages – in fact, you can read this book like you read a newspaper. There are plenty of drawings to help explain certain processes and the chapters are well delineated by subject. The intent is to communicate clearly the changes with which you are already in various stages of being familiar. The purpose of this book is to find out exactly what those changes are.

To Doris Medina

1929–1993

PART ONE

Who ages?

INTRODUCTION

Our first task in the beginning of this book is to attempt to define the process of aging. By the end, you will find this task will be mostly unaccomplished.

The reason for this ambiguity is manifold, and perhaps surprising. The problem is that there are so many ways to look at the roots of biological maturity. Some look to aging's final obligation, death, and attempt to work backwards from the event to describe what aging means. But even death can be difficult to define absolutely, in an it-makes-sense-to-the-biologist language. We will understand this ambiguity best by attempting a definition of our own. And we will do so in the same backwards style, first examining the process of death and then working our way in reverse.

A working definition

At first blush, the inability to define death, and the body's prior preoccupation with survival, might sound odd. We have no ambiguities, for example, surrounding the material facts of such notable nonagenarians as the playwright George Bernard Shaw. Not only do we know when and how he expired (at the age of 94, after suffering a 108-degree fever), we have a general idea of what happened. He *died*, and from this event a corpse and a funeral were created.

And so it is for most of us. Death appears to be a definable, monolithic, biologically irreversible fact of life. Quibbling with its consistency seems a strange exercise, even an irrelevant one. We are forced, for better or worse, to link the words 'inexorable' and 'death' in a strong bond. Such a linkage is true, however, only if you don't look too closely.

Overarching definitions run into semantic and conceptual obstacles

quite easily. Automobiles and aunts both age, for example. But then, so
do wine and cheese. We surely don't imply the same physical process –
or outcome – is occurring in each. The only commonality is a certain
time-dependent physical change, even a deterioration.

Because of this ambiguity, many researchers tether the process of aging
to an event that appears more definable, at least to biological organisms.
That event is 'natural causes'. Scientists think of aging in terms of prob-
abilities, with mounting tenure increasing the likelihood of expiration.
Aging, in their minds, is a decrease in the chance of survival. Death is a
cessation of that decrease.

Though it leaves any explicit consideration of reproduction out of the
picture, this definition of aging is not a bad start. Everything fated to
have a beginning is also doomed to have an ending. We share, along with
light bulbs and fan belts, an extinction so predictable that it almost appears
planned. Focusing on death as the end point of this planned obsolescence
gives a comforting linearity to our definition. But we deteriorate to what?
'Cessation of a decrease' has to *mean* something. Since all living things
seem to undergo it, there must be some common thread to their experience of death. We ask a single question: Is there an overarching, universal
definition of the biological process of death? This question lies at the
heart of our ability to understand the aging process in biological life. It
is the purpose of the chapters in this section to answer it.

To accomplish this task, we will first discuss the world of non-human
biology, exploring the process of death in a variety of vertebrate and
invertebrate organisms. Second, we will consider the mechanisms of
human expiration, looking both biologically and historically for an explanation of our 'moment of death.' Finally, we will explore the evolutionary
context of aging and death. By examining the force of natural selection
on senescence, we will attempt to find a biological reason for its existence.
Once a context is established, understanding the purpose may help us
also understand the substance.

A few ground rules

Several issues need clarifying before we begin. I will be using the terms
'aging' and 'senescence' interchangeably. This has its hazards, whether
we consider multi-cellular organisms or simple single cells. Botanists, for
example, use the term 'senescence' when they describe deciduous trees

shedding their leaves; that does not mean the tree is 'aging' in the traditional sense. Certain cells in our body undergo the process of senescence. This is a fairly well established series of specific internal biochemical events not necessarily leading to death, and not part of common definitions. Some cells age and die in living things – even in developing embryos – but leave the rest of the organism youthful and growing. It is important to distinguish between whole organismal aging processes and mechanisms occurring within individual cells. When considering evolutionary theory, we must distinguish between the forces exerted on an individual and those exerted on a group.

In addition to the terms 'aging' and 'senescence', I will soon be using words like 'cells' and 'cell cycle.' Although more formal definitions will follow as these pages unfold, I will give a brief description here. You remember from grade school that all human beings are composed of cells, those small, grease-lined objects that look like beach balls (Figure 1). A typical human has 60 trillion of these structures. Each cell has a nucleus, which is a glorified storage container for human genetic information. As you may recall, this information is locked up in structures called chromosomes, and is made of DNA.

In order to keep us healthy, these cells must make copies of themselves. To do that, they simply copy their genetic information and then split in two. This process, a highly controlled and very complex event, is called mitosis (Figure 2). It is regulated for a critical reason, one that can be seen when the controls break down. If the cells start proliferating in an uncontrolled fashion, a tumor may be formed.

With that bit of biology under our belts, we can return to the subject of this chapter. We are going to analyze the notion of relevance and irreversibility of senescence in living things. Placing our own undeniable human mortality in the context of the natural world may increase our appreciation of life. And give us a firmer understanding of the inevitability of our own doom. We may discover if George Bernard Shaw was correct when he joked to a comrade about the nature of his epitaph.

> I knew that if I hung around here long enough, something like this
> was bound to happen.

Animal cell architecture

A 'typical' animal cell, drawn below, looks something like a fried egg.

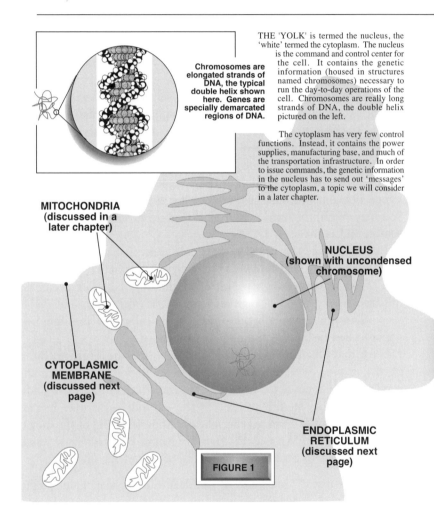

THE 'YOLK' is termed the nucleus, the 'white' termed the cytoplasm. The nucleus is the command and control center for the cell. It contains the genetic information (housed in structures named chromosomes) necessary to run the day-to-day operations of the cell. Chromosomes are really long strands of DNA, the double helix pictured on the left.

The cytoplasm has very few control functions. Instead, it contains the power supplies, manufacturing base, and much of the transportation infrastructure. In order to issue commands, the genetic information in the nucleus has to send out 'messages' to the cytoplasm, a topic we will consider in a later chapter.

Chromosomes are elongated strands of DNA, the typical double helix shown here. Genes are specially demarcated regions of DNA.

MITOCHONDRIA
(discussed in a later chapter)

NUCLEUS
(shown with uncondensed chromosome)

CYTOPLASMIC MEMBRANE
(discussed next page)

ENDOPLASMIC RETICULUM
(discussed next page)

FIGURE 1

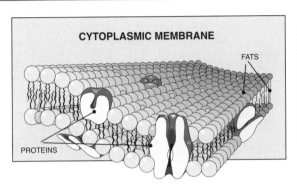

CYTOPLASMIC MEMBRANE

FATS

PROTEINS

The outer membrane is primarily composed of fats. Bobbing up and down in the membrane like a buoy are proteins. They often serve as 'receptors,' binding to external molecules and communicating the fact to the nucleus.

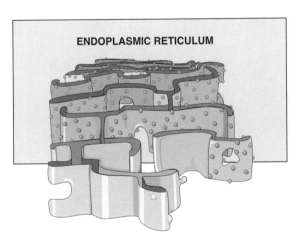

ENDOPLASMIC RETICULUM

This tongue-twisting structure can be likened to an intracellular superhighway. The ER consists entirely of membrane-enclosed spaces. These spaces provide a network of conduits where various molecules can be transported.

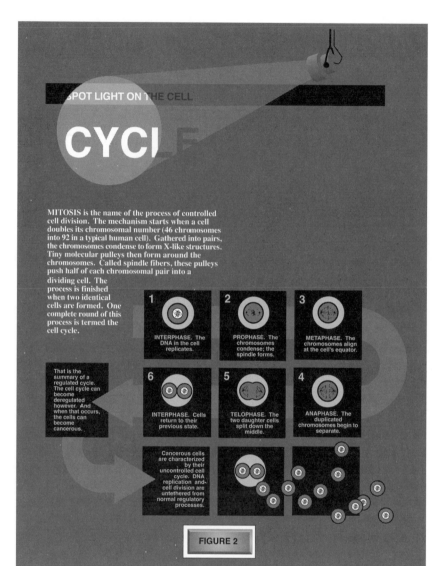

SPOT LIGHT ON THE CELL

CYCLE

MITOSIS is the name of the process of controlled cell division. The mechanism starts when a cell doubles its chromosomal number (46 chromosomes into 92 in a typical human cell). Gathered into pairs, the chromosomes condense to form X-like structures. Tiny molecular pulleys then form around the chromosomes. Called spindle fibers, these pulleys push half of each chromosomal pair into a dividing cell. The process is finished when two identical cells are formed. One complete round of this process is termed the cell cycle.

That is the summary of a regulated cycle. The cell cycle can become deregulated however. And when that occurs, the cells can become cancerous.

1 INTERPHASE. The DNA in the cell replicates.

2 PROPHASE. The chromosomes condense; the spindle forms.

3 METAPHASE. The chromosomes align at the cell's equator.

6 INTERPHASE. Cells return to their previous state.

5 TELOPHASE. The two daughter cells split down the middle.

4 ANAPHASE. The duplicated chromosomes begin to separate.

Cancerous cells are characterized by their uncontrolled cell cycle. DNA replication and cell division are untethered from normal regulatory processes.

FIGURE 2

1

A slippery overarching definition

By all accounts, the movie star Rudolph Valentino was the premier womanizer of the 1920s. H. L. Mencken once described him as 'catnip to women.' Many of the women simply called him 'jerk.'

Intelligent, stunningly handsome, and acutely aware of both characteristics, the Italian born movie star became one of the most enduring icons from the age of silent movies. He was also one of the silver screen's first male sex symbols. In the space of only five years, he had gone through three wives, countless lovers and a million broken hearts. Even so, the most startling characteristic about his career wasn't its physical intensity, but its amazingly short duration. Valentino starred in his first film at the age of 26. Five years later he was dead. As a result, this sexual supernova left to posterity only images of youth and vigor, producing an eerie timelessness that haunts the minds of many film buffs steeped in the lore of American cinema. This flexibility in our perception of aging, indeed the wobbliness of the very definition, is the focus of this chapter.

How Valentino died

The events that would take Rudolph Valentino's life started in his New York hotel suite. Witnesses relate that Valentino, lounging around his room on a lazy day in August 1926, felt a sudden, incredibly painful stabbing in his side. The pain persisted, but he refused to be hospitalized. Instead, he spent the night in agony, rolling around the floor of his hotel room. Only when his temperature soared did the semi-delirious Valentino consent to treatment.

He was taken to a nearby hospital and rushed into surgery, where the doctors found a perforated gastric ulcer. They also discovered, to their surprise, a ruptured appendix. An ugly infection had already spread throughout the peritoneum and Rudolph Valentino was immediately put on the critical list. So advanced was the deterioration that the attending physicians held little hope of recovery.

The hospitalization of America's favorite lover-boy shocked the nation. Hundreds of star-struck fans vowed to kill themselves if their hero died. Interviews with his ex-wives (as well as friends, lovers, acquaintances and anybody else who might sell papers) became the common subject of the late summer media. As Valentino's condition worsened, the publicity began to look ludicrously presidential, with press releases describing Valentino's status issued every hour. The star held on for a total of eight days, finally dying at noon on August 23. The *Daily News* led with this headline: 'The Great Director Stood Ready Today to Call Rudolph Valentino Off the Screen of Life.' After a lot of fanfare, including a cross-country railroad trip, he was buried near his home in Hollywood, California.

As dramatic as his death was, film buffs probably don't envision an over-popular corpse when they remember Rudolph Valentino. And none of them will ever recall images of the film star as an old man, although he would be a centenarian were he alive today. The reason for this is simple: Rudolph Valentino found the celluloid version of the fountain of youth. The viewing public, and the historians who studied his contribution, never got to see him grow old. Only the liquid eyes of a romantic Italian, the dashing, muted 20-something sheik wielding a scimitar remained after the grave.

Other aging movie stars have tried to achieve this form of immortalization. Doctors repeatedly alter the bodies of celebrities through plastic surgery. Editors airbrush and electronically alter their photographs. The stars hire personal fitness consultants to regulate both exercise and diet. When the inevitable deterioration comes, some aging celebrities refuse to be seen in public; their fans will thus never observe and remember an older, less glamorous image. This bartering with the decades affects the viewing public just as radically. As long as we disconnect our romantic physical memories from our harsher physical realities, we become comfortable with the fact that aging is not an inexorable process for everyone. In other words, our definition of aging has room for flexibility.

Science land

The idea that the definitions of aging and death have elasticity are not the sole domain of a movie-going public, but that fact is hardly the way we usually think about aging and death. Most of us feel comfortable with the evidential fact of *individual* death. All of us have seen animals undergo aging and death many of us have seen close friends and loved ones run through the process. The fact of its existence for all living things is inexorable, unambiguous and, from a scientific point of view, quite inaccurate.

If one looks closely at the natural world, one sees a flexibility reminiscent of Hollywood's attempts at immortality. Only this isn't the magic of a sound stage. There are many ways organisms subvert or ignore processes we humans usually think inevitable; so many ways that the ability to explain these processes using traditional definitions can be called into question.

In this chapter, we are going to explore this definitional ambiguity. I will start by discussing the notion of life span, observing organisms with tenure on the planet that varies from centuries to hours. The fact that aging and death are optional processes for many organisms will be discussed, with comparison of the biologies of creatures who must die with those who only might. Finally, problems in defining death in complex multi-cellular organisms will be examined, concluding with some thoughts I have had studying certain human cells in the laboratory – cells which are very much alive, but extracted from people who died years before I was born.

Life spans

We are very used to thinking of aging and death in concrete terms. When contemplating life we inevitably assume the presence of an internal clock. Wound to zero at birth, it incessantly and inerrantly ticks away during our entire terrestrial tenure. So solid are these concepts in our mind that we have coined the term 'life span' to denote its boundaries.

Note that I didn't use the word 'longevity' to describe these boundaries. Longevity can be defined in terms of an expectation, calculated from birth. This is the number which answers the human question, 'Given the current environment and culture, how long can a given person *expect* to live?'

A person living during the Golden Age of Greece (500 BC) could expect to live about 38 years. In the Dark Ages that figure plummeted to about 30 years, depending upon the century.

The notion of longevity can be contrasted with the idea of life span. Life span can be defined as the maximum time a person *could* live, given favorable living conditions. The idea attempts to circumvent external inputs and directly addresses the biochemical hourglass instead. Some researchers believe that the human life span is about 115 years. The exact number varies, of course. Life span is very difficult to calculate under the best of circumstances, simply because longevity can give us so much background noise. We really have no idea, for example, whether life span for humans has changed over the centuries, even though we feel fairly confident about the changes in longevity.

With these definitions in mind, we can return to a discussion of our internal biological time pieces.

No predictions here

We seem to share these mysterious biological clocks with every other living thing, and we know just as little about theirs as we do about our own. As is true for our own experience, we have found no external method to predict an individual animal's life span. The word to describe the life spans of living things is 'diversity.' For example, the golden splendor beetle (*Buprestidae aurulenta*) has a life span that can stretch more than four decades. There is a documented case of a Marion's tortoise (*Testudo sumerii*) that was accidently killed after it had lived for more than 150 years. In California, there exist sequoia pines that started growing when the Egyptians were first building their pyramids. In the same state, a creosote plant nicknamed King Clone has been estimated to have been on the planet for nearly 12 000 years.

There are also life spans that are of much shorter duration. The killifish in central Africa, for example, lives only eight days. There are insects that are born in the morning, are sexually mature in the afternoon and are dead by nightfall. The glaucoma protozoan splits from itself every three hours. In one day, the original will be a great, great, great, great, great, great grandparent to the 512 descendants now swimming within its watery domain. Many bacteria have an even shorter time, replicating every 20 minutes. These single-celled organisms possess some extraordinary biological characteristics, so far removed from our own reality that their life cycles actually impinge on our ability to define life span and 'aging.' We will return to this shortly.

Flexibility

The figures regarding life span begin to lose their monolithic character-istics when one considers individual life styles and circumstances. If an Asiatic elephant is raised in captivity, for example, you can chop more than 50 years off its life span (average death at 24 years). Chimpanzees will live less than half their life expectancy behind bars (average death at 15 years). The process is flexible enough so that you can, to a limited extent, do the reverse. Sterilize a dog, male or female, and you will extend its life by two years. The same thing can be done to a cat. A human who stops smoking increases his or her longevity by many years.

There is flexibility in the number of years different species have on the planet, but different organisms within the same species can live a very long time. For example, there is a cockatoo on record that lived for more than 80 years. There is a cow (named Modoc) that lived for 78 years. The verifiable record life span for a canine is held by an Australian cattle dog, which hung around the Outback for 29 years. The record is 34 years for a cat, and even then it did not die of natural causes, but was 'put to sleep.'

Slime and punishment

This variation can be seen even in the microscopic world. The diversity is so dramatic that the fact of death, let alone the aging process, actually becomes an option for many organisms. You did not read that wrongly. As long as a proper food supply and a steady environment are maintained, there are living creatures that *will not die*. This longevity has actually played a large role in our own survival as a species, and in this section, I'd like to talk about it.

There is an organism that possesses such an interesting name that the first time I ever heard it I laughed for an hour. The organism is a 'slime mold' (its proper name, not nearly so creative, is *Dictyostelium discoideum*) and it roams about on the floors of many forests throughout the world.

What this interesting organism actually looks like depends on which stage of the life cycle one is examining (Figure 3). There is the water balloon stage of its existence, for example. Here the slime mold resembles a bag of liquid rolling around the ground in search of its favorite food, tiny bacteria. These creatures are called myxamoebae. The organism shares many characteristics with the true amoeba, including reproduction strategies. That is, when it gets ready to create offspring, the slime mold simply doubles its genetic information, creates the biological equivalent of a fault line down its middle, and then splits into two parts.

The most extraordinary characteristic of this maternity plan, called binary fission, is its permanence. As long as the creature is fed and watered, it will never undergo an 'aging' process and die. It is content to live forever on the forest floor, happily munching on tiny bacteria, doubling in size and splitting forever. For this organism the aging process and consequent death are not mandatory, but rather optional, totally dependent not on an internal biological clock but on external environmental stability.

Sounds weird? The life cycle of the slime mold actually gets stranger, especially if the environment becomes unstable. When one of those little water balloons gets hungry and cannot find any food, it sends out a distress call to the rest of the forest floor. This summons is chemical (cyclic AMP is the formal name) and can be sensed by other slime molds creeping along the ground. When they encounter the chemical, they stop what they're doing and start migrating towards the sender. A better word would be 'streaming'; the process often involves tens of thousands of organisms, all converging toward a single center.

As busy as it is, this aggregate does not just remain a disorganized mass of streaming amoebae. Rather than just crawling over each other to form a writhing, confused pile, the creatures begin to work together in an organized fashion. An outside is created, an inside is established and the whole thing gets encased in a slimy coat. *The pile forms a single, complex, multi-cellular animal!* This latter-day Noah's ark begins to move in a deliberate and calculated fashion and even has its own name; the organism is now called a slug (or pseudoplasmodium). Once established, the slug goes searching for food and water. If it finds some, the little water balloons will abandon ship, crawling off the slug and resuming their normal life. If it doesn't find food, the slime mold will take a swan dive into the nearest mud pile, create spores for more myxamoebae, and then die.

This interesting organism is under intense investigation, because of its ability to oscillate between single-celled and multi-celled stages of life. But it also oscillates between two stages relevant to our discussion. This organism can straddle the distance between 'immortality' and certain death. The slime mold has the option, if the environment cooperates, to live permanently, reproducing by binary fission. If the environment doesn't cooperate, the organism will definitely enter a part of the life cycle that can kill it. If the slug finds food, its individual members can leave the commune and continue their great longevity.

How difficult it is to apply our overarching definition of individual death when we consider the slime mold. The ambiguity does not lie with its life cycle, but merely with our attempts to organize it. Do the individual amoebae 'die' when they aggregate to form the slug? Does the slug 'die' when the individual members crawl off and resume the water balloon stage? How can we say a creature grows old when it can flit so easily between mortal and immortal stages? Applying our human experience to such creatures can be very difficult indeed!

It might be comfortable to assume that the slime mold's life cycle is very exceptional. That way, the dilemmas posed to the creation of a fundamental definition of death can be ignored. But this extraordinary creature is not just an aberration. The slime mold is only one of many multicellular creatures that challenge our notions of life and death. To examine our next life cycle, we will exchange the leaf-strewn forest for the salty depths of the sea.

Sponges

In a minute, the subject is going to be sponges, but to illustrate their biology, I'd first like to relate a story, based on a piece of music by Paul Dukas. The story, like the music, is often called *The Sorcerer's Apprentice*, and I first heard it in the wonderful comfort of my mother's lap.

The task of the apprentice was to fill a basin with water from a nearby river. In an effort to save time, the apprentice, who was lazy, got hold of the wizard's magic book. He soon learned to cast a spell over a broom, which instantly sprang to life. The apprentice then trained the broom to pick up a bucket, take water from the river, march over to the basin and fill it up. The problem was that, when the task was finished, he did not know how to stop the broom's lively activity.

'Now, John,' Mom would say, 'the apprentice was in a fix. How

CREATURES THAT DON'T DIE

For some organisms, death is an optional part of their life-cycle

The reproductive strategies for many creatures do not include mandatory senescence. As long as there is an adequate supply of nutrients and a stable environment, existence and reproduction can occur without interruption. And without death.

Illustrated on these pages are the life cycles of four of these potentially 'immortal' creatures. They live in a variety of environments and are generally very simple organisms, like sponges and slime molds. Only some of the reproductive strategies available to these creatures are illustrated.

Bacteria in the air, soil and water

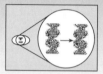
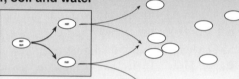

The reproductive life cycles of bacteria are similar throughout the planet, regardless of the local environment. After the DNA replicates itself, the organism splits into two. Each half receives one copy of the duplicated genetic information. After a short period of time, the process is repeated: each half duplicates its genetic information and then divides once again. Except for environmental constraints, there is no defined limit to the number of times this may occur.

Life cycle of a sponge

Sponges can reproduce sexually and asexually. If they are disaggregated, new creatures will form out of the separated clumps. This process can be repeated indefinitely.

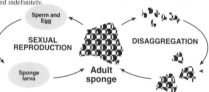

Sperm and Egg

SEXUAL REPRODUCTION

DISAGGREGATION

Sponge larva

Adult sponge

Sponges are among the simplest multicellular creatures known. Shown on the left is one type of sponge cell, called a choanocyte. It is used primarily for feeding.

FIGURE 3

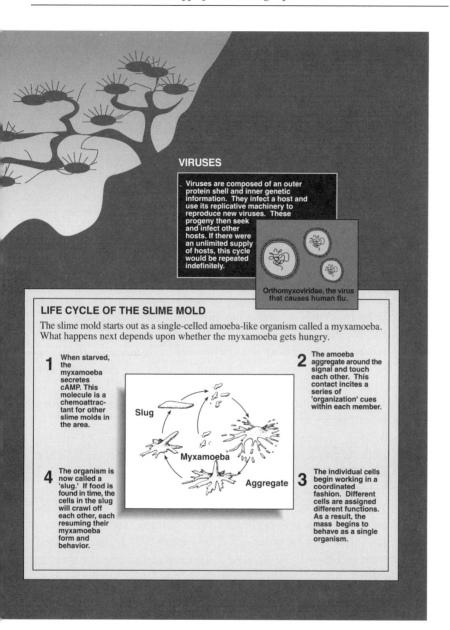

VIRUSES

Viruses are composed of an outer protein shell and inner genetic information. They infect a host and use its replicative machinery to reproduce new viruses. These progeny then seek and infect other hosts. If there were an unlimited supply of hosts, this cycle would be repeated indefinitely.

Orthomyxoviridae, the virus that causes human flu.

LIFE CYCLE OF THE SLIME MOLD

The slime mold starts out as a single-celled amoeba-like organism called a myxamoeba. What happens next depends upon whether the myxamoeba gets hungry.

1 When starved, the myxamoeba secretes cAMP. This molecule is a chemoattractant for other slime molds in the area.

2 The amoeba aggregate around the signal and touch each other. This contact incites a series of 'organization' cues within each member.

4 The organism is now called a 'slug.' If food is found in time, the cells in the slug will crawl off each other, each resuming their myxamoeba form and behavior.

3 The individual cells begin working in a coordinated fashion. Different cells are assigned different functions. As a result, the mass begins to behave as a single organism.

Slug

Myxamoeba

Aggregate

could he stop the over-enthusiastic broom from performing its chores? Spying an axe in a corner of the room, he got an idea. The apprentice quietly snuck up to the broom and in a flash, chopped it into a thousand pieces!' My eyes grew wild with excitement. Somewhere in the back of my mind I was thinking about how babies are made. My mother continued: 'To his horror, that did nothing to stop the activity. Each piece quickly transformed itself into a living broom, each marching to the river to get water. And where the poor apprentice only had one broom with which to deal, now he had a thousand!'

Mom would usually end the story as a morality play. She would describe the anger of the wizard, the penalty of laziness, and its application to a very specific 20th century little boy. And I took it all in, never realizing that such a story would have biological correlates for me one day in the world of research science. Which it does. It turns out that there are organisms in nature that can act like the apprentice's enchanted broom. One of them is the common sponge, a hardy organism that has been largely replaced by synthetic analogues in the modern household.

The sponge is interesting because, like the slime mold, it is really only a loose aggregate of cells. There is a division of labor amongst them, but there are no systems, no organs, no mouth or digestive tract, and only primitive neural integration. This rather loose coalition is being studied because, like the slime mold, it represents one of the simplest organismal structures in nature.

From our point of view, the most interesting characteristic of sponges is their viability. If you have the right chemicals, the sponge can be disaggregated and reduced to its component cells, no organization allowed. If that process were applied to you and me, that act would be a death sentence. For the sponge, the violence is merely another way to reproduce. These disembodied cells will eventually form clumps, and with time, a familiar organization will be observed. The sponge will regenerate itself. Like the broom in Dukas' *The Sorcerer's Apprentice*, dividing the creature is merely another way of multiplying it.

This puts a slant on our discussion of death and aging in the following way: Each individual cell of the sponge carries some very important information, summarized as follows: 'I know how to reconstruct myself completely.'

The problem with applying a fundamental definition of individual death to a sponge has to do with numbers. How many organisms does it *really* take to make up a sponge? When you disaggregate a single large sponge, do you really 'kill' a conspicuous organism, or simply throw it into a new stage of reproduction? Is this just a semantic exercise? Does the concept of death really apply here, or, as in the slime mold, do we only see optional demise?

Once again our definitions of death are going to have to take on a new form of flexibility. The concept isn't completely irrelevant with these creatures; you can start The Clock of Ages within slime molds simply by starving them at the slug stage. The point here is simply that the biological idea of aging and death is flexible, and there are organisms that have evolved around the process in order to survive.

More extraordinary still, there are other organisms that *never* carry a clock with them. They are so simple and so small that millions could fit very comfortably on the period at the end of this sentence. Don't let their small size fool you, however. As a result of their life cycles, they have changed the course of human history.

So what?

One of the organisms responsible for dramatic changes in human politics and behavior is a bacterium called *Yersinia pestis* (also termed *Pasteurella*). As its name implies, this small rod-shaped organism is responsible for the bubonic plague, a disease that has killed many millions in human history. It has been able to do such genocide because of a unique aspect of its life cycle. Like almost every other bacterium on the planet, *Yersinia* never 'ages,' at least in the traditional sense of the word. This bug's ability to live its rather continuous life in large, mammalian creatures is the reason that it was and is so deadly. This is how it works:

Yersinia pestis is a parasite of various rodents, living part of its life cycle in the blood. The bacterium can be transferred from one rat to another by the blood-sucking flea. *Yersinia* proliferates easily in the guts of these insects, replicating itself so dramatically that it actually blocks the flea's 'throat', preventing the flea from getting nutrients. The flea becomes terminally ravenous. When the starving insect bites its victim, the blood simply goes down its throat and mixes with the growing *Yersinia* colony. Because there is blockage, the now contaminated blood is regurgitated back into the wound the flea has just created. As a result, an innocent creature is infected. The flea might as well be carrying a syringe full of plague.

How does *Yersinia* proliferate so rapidly? Why doesn't it just grow old and die? The reason has to do with its rather boring sex life. *Yersinia* doubles its genetic information about every half hour or so. It then divides into two organisms, stuffing half of the information into each 'new' self. Many minutes later, each half doubles its genetic information and then repeats the process. If given the proper nutrients and a stable environment, *Yersinia* will do this forever. There is no aging of this organism, there is no 'life span' that causes it inexorably to roll over and die. It's more like a perpetual motion machine from hell, living off its environment, creating a destructive force that started in antiquity and continues to kill today. Death for this organism is thus not a requirement. It is simply an option.

The numbers

The idea that a creature might really look at death only as one option in its life cycle might seem rather intuitively strange. But if it were known that most organisms look at death that way, these facts might seem like a lie. They are not lies, of course. There are probably more bacteria in a 10 pound piece of rotting cheese than there are all other organisms on the face of the earth. And since the other bacteria follow *Yersinia*'s replicative model, we must conclude that most creatures on the face of the planet don't view aging and death as mandatory. The bugs all have a defined 'life span,' of course. Give a bacterium like *E. coli* 20 minutes and it gets ready to divide again. But here's the creative loophole: the parent does not eventually die, simply because it has an offspring. It subverts the model by *becoming* the offspring, at least in part. This ensures that some part of the original can stay on the planet, happily rotting people or cheese, for centuries.

The definitions of aging and death get harder to apply with simpler and simpler organisms. Once we get to some of the least complex creatures on the planet, the definitions seem to disappear altogether. Viruses, for example, are nothing more than bits of protein that have been stuffed with genetic information. Because of their chemical structures, many viruses can actually be crystallized. Just like certain rocks. And when they are crystallized, they can lie dormant for literally thousands of years, become resurrected in a drop of water and restart their life cycles as if no time had passed. It is very difficult to define the aging process in a creature that has as much in common with geology as biology.

Death decisions

We have been using the examples of several odd organisms to illustrate the idea that aging and death may be more flexible concepts than we might first have thought. But does that have any relevance to more complex creatures such as ourselves? If it is true that most organisms don't go through the normal life cycle that is foisted upon us humans, where is the great dividing line? Is there some kind of rule that separates those who may sign up for death as an elective from those who must sign up as a requirement? There are many researchers who think there is such a law. To discuss it, I'd like to tell you about a salmon dinner I had recently with a fish biologist friend of mine.

The dinner

I have to admit I wasn't very hungry. But the restaurant in which we met was situated in a beautiful part of Seattle, right on the water, the reflection of the skyline waving like metal and chrome seaweed in a calm Puget Sound. He sat down and ordered a new appetizer the restaurant was trying, little slices of octopus legs. I declined his invitation for a taste, ordered my dinner and waited.

The server soon brought out this delicious pink salmon filet. My friend, who ordered a halibut in a kind of lemon sauce, was equally enthusiastic. The difference in our orders immediately led us, ever the scientists, to a discussion about the biologies of the two fish. He asked about their genetics and I asked about their life cycles. How is it that these come back to spawn? And why do they die after they mate?

'This meal reminds me of a trip I took as a graduate student,' my fish biologist friend said. I could tell he was getting ready to tell a long story. My friend put his napkin to his mouth and began to talk not about salmon, but about his appetizer: 'The goal of that trip was to find out more about the mating habits of the octopus,' he continued. 'You know that the biggest ones on earth are probably in these waters.' He was right, of course. The largest ones ever observed were found in Puget Sound. 'They're pretty intelligent animals. But they lead a pretty sad life in the end, especially the female.' He smiled. 'They only mate once, you know.'

He then went on to describe the fact that after an octopus mates, the female takes the fertilized eggs and attaches them to various ledges in her grotto. 'The place looks just like a grape arbor,' he related, 'thousands and thousands of eggs.' After the female ornaments her watery apartment with her living decorations, she will take in no more food. Instead, every few seconds, she will blow water over her egg clusters. This helps to keep them clean and may aid in development. She does this literally for weeks, growing steadily weaker and weaker from lack of nutrition. Finally the tiny eggs hatch.

'Most of the little guys don't make it,' he said. 'They're fully formed and ready to float, but . . .' And then he snapped his fingers. 'That's okay though. If you think about it, they only need two from those many thousands to maintain current population levels.'

And so it is. The mother octopus, her protective job now over, is totally exhausted. Soon after the hatching occurs, she dies. 'It's one of the saddest sights in all of biology,' my friend concluded.

The rule

And what exactly does the mating characteristics of the salmon and the octopus have to do with deciding mortality amongst terrestial creatures? The rule does not so much have to do with *how* the mating occurs, but with the fact *that* the mating occurs. The rule, simply stated, is this: If you have sex, you will eventually die.

If you are male and female, you will have to mate in order to create offspring. And if you have to mate, your expiration is assured. To understand how this all works, we will leave this 20th century salmon dinner and describe a little earth prehistory. We will wind the clock back to earth circa 3–4 billion years BC.

The origins

It was not a friendly place, even if you had a life vest; it was certainly no real home for complex organisms. In this soup, most of the living creatures resembled in reproductive style the bug that causes the plague. These primitive organisms would double their genetic information and just split into two. No discarded corpse, no pining sons and daughters.

This style was probably *de rigueur* for 2 billion years. It is, as we have seen, still in vogue now.

But reproductive styles, like any fashion worthy of the name, can change. The have-sex-and-die model was very much unlike the just-split-and-live one. Here, the parent did not become one of the offspring. The parents just stayed parents – making a third organism from pieces of their own constituents. That young organism could grow up eventually to do the same reproductive trick with a compatible mate. The parents, their contribution made, would just die. This strategy not only produced a third character, but also introduced a funeral, a mandatory, tell-tale corpse. And as there was a definite reproductive landmark, time could be measured in an organism as pre-reproductive, reproductive or post-reproductive. This produced the concept of aging.

Although no one really knows, sex was thought to originate about a billion years ago, possibly coming as a response to that other great biological drive – eating. Creatures may have devoured some of their own kind, in some cases incompletely digesting the victim's nucleus. The satiated creature's nucleus may have fused with its victim's, doubling its genetic information and perhaps conferring a selective advantage. In this fashion, the mixing and matching of genes could increase survival chances in unstable environments by increasing the variety of the creatures that could endure them. Indeed, organisms that don't have sex tend to be small and look like tiny pills under the microscope. Organisms that do have sex range from palm trees to people.

The bottom line is that aging followed by death became the terrible price we paid for sexual reproduction. The idea was to increase the variety and thus survivability of the species in an unstable environment; this purpose would be countermanded if the mature citizens were allowed to hang around as long as their children, especially if there came a time when the parents were no longer reproductively fit. Like some dashing shiek in a Hollywood movie, sex came on to the scene. And running alongside it, so did death.

The complexity grows

This idea of shared genetic material bringing about mortality also brings about ambiguities in our attempts to establish an overarching definition of aging and death. Consider the fact that there exist creatures who carry other creatures in their bodies normally, vitally and without infection. We will examine how this arrangement challenges commonly accepted ideas about aging and death by discussing the interesting relationships between certain animals and their biochemistries. We start with an organism that likes trees for breakfast.

Termites

Most of you are familiar with termites, those extraordinary insects capable of devouring massive quantities of wood. You may be less familiar with the fact that termites are not capable of extracting nutrients from a single particle of the lumber they consume. They certainly take bites out of the wood, but without a curious organism existing in the gut of the insect, termites would all die from a rather nasty kind of constipation.

The extraordinary organism existing in the termite's gut is called a trychonymph. Under the microscope, the trychonymph looks like a hairy beachball. This organism possesses the ability to blast pieces of wood into molecular smithereens, which it does whenever the termite eats and swallows. Once the pieces of wood are broken down into more compatible products, the termite can reabsorb them for fuel. In return, the trychonymph is provided with a very comfortable home. This is an example of biological cooperation, a process called 'commensalism' in science land.

The relationship between a trychonymph and a termite is an agreement etched in mortality. It gives us an example of the complex nature of death. If for some reason a group of trychonymphs all of a sudden die in a termite gut, the termite will die too. Not because the termite *really* expired; its biological clock may have given the insect a substantial longevity. It's just that the well-being of a termite is very dependent on the mortality of its passengers. The same thing can be said if the termite expires before the trychonymphs die. The little inhabitants are doomed. In fact, a trychonymph life span can be measured in terms of the survivability of its host.

What does individual death really mean when one considers this agreement? In many ways, it depends on your point of view. Either organism can die and, in so doing, pass a death sentence on the other. The other may not be ready to die, may in fact be able to live a long and productive life with the proper partner. But that lack of a partner sets the other on an inexorable path. This example serves to illustrate the complexities of dealing with multi-cellular creatures. It does it by expanding our definition of 'organism.'

A lesson for us

The example of a trychonymph actually has some relevance to human biology. Our own bodies are made of complicated systems, which are in turn made of complicated cells. 'Natural' death for most of us occurs not because of the sudden failure of all 60 trillion cells, but because a critical component stops working. In mutually dependent multi-organ systems, the breakdown of a single component can result in the destruction of the whole. That is why it is difficult to say that someone died of 'natural causes,' which implies the equally ambiguous term 'death by aging.' Once again we see that our traditional notions of aging and death will need some review when multi-cellular creatures are considered.

Our bodies have associations very reminiscent of that of trychonymphs and termites. Structures exist within our cells that may have at one time been living bacteria, lurking around our cytoplasms like cellular octopi. These structures, called mitochondria, work like tiny batteries, providing fuel and energy for our various needs. It is thought they were once free-living bacteria because of their structure; they have outer walls like bacteria and even possess genetic information, tied into a circle like that of bacteria. Long ago our progenitor cells captured them and began using their unique talents for their own purposes.

So what happens if these mitochondria are functionally destroyed? The cells that carry them will have no energy resources and fairly soon they will die. How do we know that? There are a number of poisons that work by stopping the ability of these batteries to perform their jobs (carbon monoxide poisoning, for example). Death occurs not because the entire organism malfunctions, but simply because it can no longer find fuel.

The point

All these variables are mentioned to emphasize a single point, and then ask some questions. The take-home lesson is that when you try to talk about aging and death in complex, multi-cellular organisms, you have to specify definitions. At what point do you say that something is dying? If the plague is treated soon enough, the infected human doesn't die; rather, he or she is simply cured of unpleasant passengers. But cure a termite of its trychonymphs, and the termite will starve to death. Talking about individual life spans is thus a function not of some organism-wide trait but of its component parts. You ask the deeper question: what *part* of the system failed?

If the multi-cellularity of organisms can complicate our concepts of mandatory aging and death, there are cellular processes that almost make them irrelevant. I cannot leave this chapter on the definitions of death and aging without commenting on at least one of these processes. In this last section I would like to talk about cancer, and the troubling role it plays in our attempts to reach an overarching definition of aging and death.

Under the hood

We begin by looking not at a single organism, but rather at single cells within an organism, and at a particular technology we use to do experiments on those cells.

It was found a number of years ago that many types of cells, including human cells, can survive outside the body. They are removed from an organ, placed into a petri dish, and bathed with special concoctions of fluid. When the cells are placed into an incubator with a very specific atmosphere and an even more specific temperature, they will grow and divide. This process is called tissue culture, or, more properly, cell culture, and has allowed us to study cellular behavior unimpeded by the donor.

The interesting characteristic of cell culture is the length of cellular survival in the dish. Different cell types can survive for different periods of time. Some can live in the dishes for many months, some only for a few weeks. As the cells get older, they lose certain familiar characteristics and gain other unfamiliar ones, making some experiments difficult to interpret. Most exhibit an 'aging' process with increasing passage of time. After a while they will even refuse to replicate themselves, almost as if they had undergone some microscopic menopause. Such 'aging' has a formal term. We call it replicative senescence. These cells don't die immediately; in fact, a majority of them remain biochemically active for a long period of time. They've just stopped reproducing. Such an interesting state is properly reflective of many cells within our own bodies. Exactly how these cells survive, and what they accomplish while living, is under intense scrutiny.

Cell replication and aging

The intuitive expectation is that, when a cell loses its reproductive capacity, it is liable to die. Then, as an organism's cells begin to lose their functions, the creature's life can be in jeopardy. The capacity to replace and/or regenerate such ailing building blocks might seem to become the primary determinant of the organism's longevity. This can be seen throughout the animal kingdom. On one hand you have small species such as adult insects and worms, which possess cells completely devoid of replication. On the other hand you have organisms like sponges, which as we have seen possess cells so reproductively active that the creatures truly have indefinite life spans. We humans, along with all mammals, exhibit gradual 'aging' with some cells capable of regeneration and others, through time, gradually losing their functions.

Because complex creatures such as ourselves possess such a mixed bag of cells, the link between aging and the cell cycle is not always clear. Despite some intense effort, ambiguity remains in our attempt to fully understand their association. For example, there is a great deal of evidence demonstrating that when cell replication ceases, the risk of overall organismal mortality increases. This occurs primarily because post-replicative cells suffer cumulative damage in their long tenure. The aggregate effect is to weaken the tissue in which the cells reside. Many organisms that rapidly senesce or are short-lived possess very few actively dividing cells.

The ambiguity lies in the fact that this association between infrequent replication and organismal degeneracy doesn't happen all the time – and our bodies are prime examples. We possess cells that replicate exactly once when we are children and never do so again. Yet they can maintain a healthy vitality through more than a century of robust living. This vigor can be seen in the laboratory merely by placing some cells in a dish under conditions where they no longer divide. They remain viable and metabolically active for long periods of time, probably reflecting their behavior in their natural biological environment.

Even in cells that are popularly used to study senescence (for example, a cell type called a diploid fibroblast), many biochemical functions are retained while the cell experiences 'old age' and eventually death.

From these examples, it is easy to understand why replication is not necessarily associated with any formal definition of senescence. It is also easy to see why we can have a hard time defining senescence in complex, multi-compartmentalized environs like a cell.

As we understand more about cellular growth, more and more types of cells are beginning to fill our incubators. Even though we can't always associate replication with aging, there appears to be a common fate. Even those cells that can grow in culture a long time before undergoing senescence appear to have a more or less finite life span. After a while, they die. We can take them out of their petri dishes, add new fluid to the cells and place them back into their incubators. We call this event a 'passage', and the manipulation seems to add to the life spans of some of the cells. But passaging the cells only stalls the inevitable for most cell types. It is true of every normal cell type currently grown that, given enough time, death will occur. Every cell type except one.

A cancer cell.

Cellular immortality

You might find it hard to believe that there are cells that don't have to die, almost as if they were paying attention to the same reproductive instructions etched within the genes of a common bacteria. But there is such a thing. If we were to remove some cells from specific types of tumors, place them into culture, and wait six months, we would find an extraordinary thing. The cells would still be around, growing and dividing like youngsters. We could change the cellular media and wait another six months; they would still be growing at the end of that period. We could even wait six years, and as long as we fed them and kept them warm, they would still be growing. I have worked with cells taken from a woman named Helen who died *three years before I was born*. Her cells are still so robust that we had to tuck all the other cells in the lab away in their incubators before we pulled Helen's out; if they landed in the other's dishes, Helen's cells would have contaminated other cultures, rendering any experiment useless.

This amazing biological process has a term. We say that the cells have been immortalized. You did not read that wrongly. There exist cells in laboratories around the world that, if properly maintained, will never die. The formal, scientific name for this condition is, no kidding, immortalization.

Curiously, we really don't even have to wait for a tumor to get an immortalized cell. There are viruses, for example, that can cause tumors. If we put 'normal' cells into a dish, we can drip a solution containing such viruses, or even parts of the viruses, into their growth media. Under certain conditions, the cells will become 'cancerous'. And lo and behold, they also become immortalized.

This is the ultimate definition of flexibility. Here we have cells normally under a death sentence, easily persuaded by a cancerous process or a few viral genes to loose themselves from the surly bonds of mortality. We also have an experimental model that allows us to ask a great queston. What is the difference between cells that must die in culture and immortalized cells that don't die at all? In subsequent chapters, we will attempt to answer parts of that question. In this chapter, we need only understand that such a question actually exists.

Conclusions

We have normally viewed aging and death as fixed processes. Their defintions seem obvious, their tracks easily observed, their consequences final. In this section, we have observed this to be true only as long as we do not look too closely. The purpose of starting out in this fashion has been to rethink the way we view aging and death biologically.

We began our discussion with the flexibility of life spans. A comparison showed that, depending on the organism, terrestrial tenure can vary from minutes and hours to decades and centuries. Even inside a single species, life span can be dictated by something as seemingly arbitrary as confinement. There is therefore no way to predict the exact number of years a particular individual, unimpeded by disease and uninjured by accident, will normally have on the planet. We can only come up with average life span, grieving when life is cut short, marvelling when it is expanded, satisfied when it is about right.

We also discussed the fact that death isn't a mandatory requirement for every living thing. When given the proper environmental conditions, there are some organisms that don't die. Or that can change form and function so dramatically that defining an aging process is almost an irrelevant exercise. We even discovered a reproductive rule: those organisms that have sex generally leave a parental corpse; those creatures that don't have sex are compensated by potential immortality.

Next we examined the problems of defining death in multi-cellular organisms. Using the example of termites, we showed that some organisms have power over the life span of others merely because of critical associations. Defining a single biological clock for such organisms is rendered difficult by the presence of such interactions. Those associations can become so intimate that, like mitochondria, the passenger actually becomes part of the overall organism.

Finally we used the illustration of cancer to show the ultimate flexibility – that is, the astonishing ability to reverse parts of the aging process. A wonderful experimental model is created by simply comparing cells that must die with cells that don't have to. The interesting point is that the model can be created out of the *same* cells. We are only just understanding the ramifications of such flexibility.

In many ways, this roto-tilling of our aging ideas can seem like a contradiction: how can aging and death be so naturally flexible when, at least for humans, it is so inexorable? It may help to remember that science only describes what exists, not what we imagine is really out there. It may also help to recall some of the examples mentioned in this section. Or perhaps remember our friend Rudolph Valentino – now forever etched in our minds as a young man, simply because his death, in its premature form, was so very certain.

2

Humanizing aging and death

You have to admit, it's quite a love line:

> I wish to believe in immortality – I wish to live with you forever.

These words were penned by a love-stricken English poet, the erudite and youthful John Keats. The target of this articulate affection was Fanny Brawne, a genteel woman raised in wealthy pre-industrial London. As sweet as his sentiments may have been, Keats didn't live long enough to give her or us a happy ending. The reason was tragically biological. Keats' lungs were laboring under the occupation of a few billion *Mycobacterium tuberculosis* bacteria, the causative agent of tuberculosis. The poet may have contracted the disease from his family, having watched both his mother and his older brother die of 'consumption.' Eventually, the disease involved his own life, abridging it to a mere 26 years.

Keats had a professional as well as personal interest in tuberculosis. Before he was a poet, he was a physician, trained at the famous Guy's Hospital in England. Keats did not turn to professional writing until the final three years of his life, inspired in part by Brawne, in part by his familial losses. Those three years would be filled with a biological wrestling match between his brain and the bacteria, between his desire to obtain immortality in writing, and his eventual desire for suicide.

Keats first suspected he might have contracted the dreaded disease during a visit to Scotland. In the summer of 1818 the poet caught a cold while vacationing on an island in the Hebrides. The cold would not go away. Keats was familiar enough with consumption to understand what a constant sore throat and fever usually meant to someone who had twice been exposed to tuberculosis.

A year passed, his constitution steadily weakening. He vacationed on the Isle of Wight, forging his poetry in the daylight hours and corresponding to his friends at night. He wrote 'I find my body too weak to support me to the height.' The soreness in his throat seemed to abate with proper rest and relaxation, but would easily return when the weather became cold or the poet over-exerted. A disturbing tightness in the chest told him his lungs were under siege. He began to dread coughing, always worried that the handkerchief he put to his mouth would come back drenched with blood. Then he would know that the waves of bacteria pounding against his lungs were at last eroding them – and him – into an uncomfortable mortality.

The symptom he had been dreading occurred almost a year before he died. 'Bring me a candle!' he shouted to a companion after a particularly severe hacking cough. He examined his handkerchief and declared 'It is arterial blood. I cannot be deceived . . .That drop of blood is my death warrant. I must die.'

His health and his attitude deteriorated together. Though Fanny Brawne was now his fiancée, he became convinced she was seeing another man. He experienced delusions, blaming the disease one week on his literary critics, the next week on Fanny's infidelity. He wrote vehemently and venomously to her and to his public.

As his situation (and disposition) worsened, Keats was advised to travel to the warmth and dryness of Italy. Accompanied by his friend Joseph Severn, John Keats set sail in September 1820. He said to his friend, 'How long is this posthumous life of mine to last?' Unknown, but perhaps expected by him, the answer would be only a few months away. John Keats would never return to England.

From a medical point of view, the trip was a disaster. There were storms and fog and dank surroundings; on board he was coughing up blood every few hours. Keats and his escort had to wait in the ship ten days just off Naples, because of a typhoid quarantine. The feverishness and delirium became constant, and it seemed he had very little lung tissue left to expectorate. Indeed, what was left of his lungs at autopsy contained virtually no unaffected tissue. He became suicidal, and knives and toxic substances on board had to be removed from his presence.

Eventually the end drew near. Sensing this immediacy, Keats began to make burial plans. He requested his grave marker be inscribed with 'Here lies one whose name was writ in water.' The words were a stunning reversal from the days when he wished for immortality so that he might fulfil his life in his love.

John Keats died in his companion's arms, February 23, 1821, and was buried in the Protestant Cemetary in Rome without fanfare.

The chapter before us

Tuberculosis was the scourge of Keats' century. The bacterium ravaged Europe because of its ability to invade in mass numbers, subvert and ultimately ignore the human immune system. As evidenced by feverish delusions and delirium, the secondary effects of the tuberculosis affected even the nervous system. Confronting this horror without even the skills of rudimentary bacteriology, the medical community was ineffective and sometimes quite harmful. The treatments prescribed during infection – bleedings, salts, enemas, even travel – did little but weaken the patient and hasten death.

In this chapter, we are going to use aspects of Keat's symptoms to extend our discussion of the definition of aging and death, this time focusing on human beings. We will use immunological, neurological and historical examples in an attempt to grapple with this surprisingly elusive concept. In the previous pages, we considered how various organisms' life cycles confounded our ability to discern a consistent, common-sense definition of individual expiration. Now we will examine just how the life cycles of human beings can do the same thing.

To accomplish this task, we will discuss human death from two directons, starting first from our cells and working outward. We will examine the kinds of death available to human cells, considering the fact that human embryo construction involves getting rid of old cells as much as creating new ones. Then we will go in the opposite direction, starting with processes already established as adults and working inward. We will discuss the ambiguities of defining the moment of death, especially a definition applicable to all cells within a particular individual. This we will explore both biologically and culturally. By examining how other civilizations and groups have grappled with the idea of human aging and death, we will discover just how many variations there are to the concept.

The hope is to extend the idea that definitions of aging and death are flexible for many organisms, even humans. To talk about cells, we will undoubtedly need to brush up on our biology. To talk about cultures, we will need to brush up on our history. Altogether, the common humanity that courses through both will give us a hint as to why we have thought the *idea* of death so immortal, and so all-consuming. In the end even Keats surrendered to its power, its inevitability, even its gentle ability to relieve chronic suffering. He wrote

Verse, Fame and Beauty are intense indeed
But death intenser – Death is life's high mead.

The need for a guard

There is a very uncomfortable fact concerning our relationship with the microbial world. I would like to use this uncomfortable fact to inaugurate our quest for a definition of human death.

At this moment you have micro-organisms in your body that would love to eat you for dinner. They are in your mouth, in your gut and on the surface of your skin. As you might expect, most of them can be pretty nasty, ranging from viruses that can give you colds to bacteria that can give you strep throat. If any of them reached your bloodstream and met no resistance, you would be dead and then you would be lunch. These organisms are such a part of your body that they are termed 'normal flora' (Figure 4).

Uneasily, humans have been able to harness some of these nasty guys for work. For example, you can read this without running to the toilet every three minutes because bacteria work in your large intestine to solidify the bowel. Without them, dramatic and chronic diarrhea would be the result. It is a troubling truce, however. If a large number of those organisms breached their gastrointestinal prison, an infection would result and you could very easily die.

We survive in this microbiological cold war because of our immune system. Human immunity consists of a complex series of cells that together possess the ability to identify who is a foreigner and who is a friend. Immunologists have long known that in order to do this, the immune system has to be able to separate self (which is friendly) from non-self (which is not). If a foreigner is identified, the system then has to visit destruction upon the invader. It's quite a job. Since invasions occur with every breath and swallow, the immune cells have to be under constant alert, apprehending *everything*, discerning the neutral from the nasty, the friend from the foe. For a long time, the question was: how do they do it?

How guards work

The modern techniques of molecular biology have begun to shed light on this critical question. We have learned, for example, that our immune

NORMAL HUMAN FLORA

The surprising bacteria that inhabit our bodies from birth to death.

Bacteria normally populate our human bodies, conferring upon us specific biological characteristics. Some bacteria perform critical functions. Others are simply along for the ride. The number of individual organisms depends upon the location of the body being examined. Some places, like the stomach, are virtually sterile. Other areas, like the mouth, have millions of bacteria per square centimeter. Researchers believe that the large numbers may serve as biological protection. Externally pathogens may never get a foothold if there is too much competition for nutrients in a given area. Listed below are the names and locations of some of these bacteria.

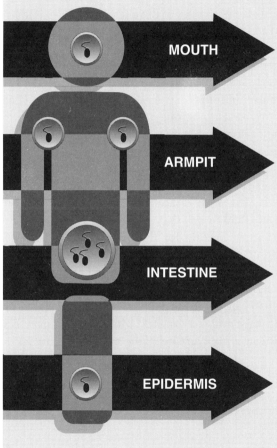

MOUTH

The human oral cavity contains one of the greatest concentrations of normal flora in the body. The 'morning breath' with which most of us awake is derived from their metabolic by-products. Chief among these are species from the genus *Streptococcus*. One such organism, *Streptococcus mutans*, is responsible for dental caries.

ARMPIT

On average, over 10 million micro-organisms exist in one square centimeter of the human armpit. The presence of salt, moisture and fatty acids creates a miniature ecosystem where a variety of organisms may thrive. This fact can be detected by the presence of body odor. Such smell, like the morning breath, is derived from their metabolic by-products.

INTESTINE

Whereas the small intestine is relatively sterile, the large intestine is heavily populated with micro-organisms. Some bacteria, such as *E. coli*, play a large role in solidifying the bowel. Other organisms synthesize useful vitamins, such as B-12, folic acid and vitamin K. In some animals, such synthesis is vital for survival. Except for poorly nourished individuals, this is of doubtful human importance.

EPIDERMIS

There are three main groups that colonize human skin. One such organism, *Propionibacterium acnes*, is associated with severe acne. Another group is from the genus *Staphylococcus*. These organisms secrete substances which are anti-bacterial in nature. They reduce the frequency of colonization by otherwise hazardous skin pathogens.

FIGURE 4

systems know how to identify self from non-self in the same way that soldiers decide who is friend and who is foe on the battlefield. They *look*.

In a war situation, soldiers examine outer apparel. If two soldiers see the same uniform, they identify themselves as friendly – even if they don't know each other. This sometimes can be deceptive. During World War II, there were rumors that German soldiers in the Battle of the Bulge were dressing up in American uniforms. This made the going very rough in some places, producing distrust in some of the allied units. The importance of possessing immediately identifiable markings during conflict situations cannot be underestimated.

In the battle to keep you healthy, your immune system does the same thing as the soldiers. Nearly every cell in your body has some kind of identifying mark, a molecular uniform, that it wears on its outside. This uniform is as individual to you as your fingerprint. You get half of the identifying marks from your Mom and half from your Dad. Thus families share similar molecular identity – and identical twins have exactly identical uniforms.

The immune system uses these identity markings as an alarm system. You have cells in your body that work 24 hours a day, doing nothing but inspecting every bit of tissue, fluid or cell they encounter. They are examining these molecular uniforms, looking for the presence of some nasty invader. If a cell or group of molecules looks familiar, they are allowed to pass unharmed. But if the surveillance cells detect something foreign, they broadcast molecular signals that can bring the equivalent of a D-day invasion onto the biochemistry of the invader. That is why you stay healthy for so long and it is why you get so sick when an organism like *Mycobacterium tuberculosis* learns to overwhelm the system (Figure 5).

How the system learns about this

The establishment of this marvellous mechanism of human immunity has a great deal to do with our discussion on human aging. To understand the connection, we first have to find out how it develops.

A little embryo does not have an immune system in its initial stages, of course. And once it acquires the cells that will eventually become the immune system, it still does not distinguish friend from foe. These cells have to be taught three obvious facts.

Fighting Foreigners

*The immune system is the best defense against unwelcome microbes.
Here's how it recognizes foreign invaders.*

Identifying and destroying harmful pathogens requires the interaction of several kinds of cells. The task is two-fold; the immune system must recognize the foreigners and then antibodies, our body's best line of defense, must be created. Listed below are the names and descriptions of three cells involved in the process of alarm and attack.

MACROPHAGE

Literally 'big eater.' Macrophages recognize foreign organisms and eat them. Some of the foreign particulate matter is regurgitated to the macrophage surface.

B CELL

These cells make antibodies. Normally 'silent,' they must be chemically stimulated before antibody production takes place.

T CELL

One of many types, the T cell listed here turns on B cell antibody synthesis. It too must be stimulated before interaction with B cells.

HOW THEY INTERACT

As stated in the text, there are cells which inspect our tissues for the presence of foreigners. One such cell is the macrophage. The process of pathogen-defense begins with the recognition of an invader.

FIGURE 5

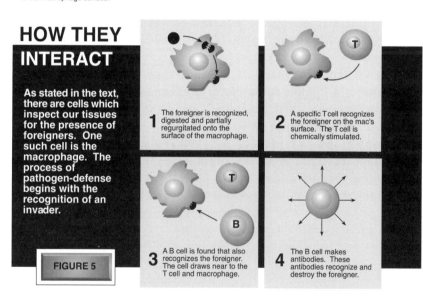

1 The foreigner is recognized, digested and partially regurgitated onto the surface of the macrophage.

2 A specific T cell recognizes the foreigner on the mac's surface. The T cell is chemically stimulated.

3 A B cell is found that also recognizes the foreigner. The cell draws near to the T cell and macrophage.

4 The B cell makes antibodies. These antibodies recognize and destroy the foreigner.

(1) Self and non-self uniforms exist;
(2) Cells that possess non-self uniforms are to be destroyed;
(3) Cells that possess self uniforms are to be left alone.

Programmed obsolescence

There comes a time early in your embryonic life when you manufacture immune cells. You make cells that carry their brand new uniforms and you make cells that keep everything under surveillance. But the system hasn't been fine-tuned. You certainly are making surveillance cells that can bind and destroy cells that don't have your uniform. Unfortunately, you are also making cells that can bind and destroy things that *possess your uniform*.

This is a very unhappy state of affairs. If your own immune system never learns to leave your uniforms alone, it will attack your cells as if they were foreign. The result? You will be dead before you are born. Obviously, these familiar yet hostile cells have to be eliminated, or at least shut down.

During development, there is a period of time when any immune cell that recognizes a friendly self uniform is eliminated or deactivated. It is a disturbing genocide, with many millions of these cells killed or made inoperative. This process is deliberate and can be defined as a kind of aging (more on that in a minute). It even has a special name, apoptosis, coming from the Greek word describing a tree losing its leaves (Figure 6). The suicide occurs in response to signals we don't understand at all. We just know it happens. The result is that our bodies learn not to destroy our own tissues. We can live a happy and violent life, throwing off microbial invaders every second that we live.

The point

This deliberately programmed cellular suicide is a strange turn of events. When it is considered how babies are constructed in the womb, most people think it is the time when young cells are made to grow and develop, not made to age and die. Yet in our immune system, dying is exactly what happens. One can see the power of this process by considering when

The process of

APOPTOSIS

Literally, falling leaves, apoptosis is a form of programmed cell death.
The events cells undergo are described below.

Cell death is a normal part of a developing organism's life, including ours. Our immune and nervous systems undergo massive rounds of apoptosis in their construction.

Unlike other kinds of cell death, apoptosis is deliberately programmed into cells. Genes responsible literally tell the cell to 'commit suicide.' Illustrated below is the process.

1 The cell's chromatin begins to condense. Chromatin is the complex mixture of chromosomes and proteins inside the nucleus.

2 The outside membrane of the cell loses its structural integrity. Protuberances called 'blebs' are observed on the surface surrounding the cell.

3 The physical size of the cell shrinks and the DNA becomes fragmented. Eventually the cell dies. Some of the genes that govern this deliberate death have been isolated.

FIGURE 6

it goes wrong; certain autoimmune diseases cause damage because some cell was allowed to recognize self and start an immunological civil war.

But are all cells actually programmed to die in an embryo? The answer of course is no. Yet to integrate the process of apoptosis – this programmed death – into our thinking about aging and dying, we see already there are several kinds of death. To contrast this deliberate cellular immolation with 'normal' cell death, consider the following childhood memory of a playground accident.

It happened to me when I was ten years old. I was playing on a wooden teeter totter with a friend when my hand slipped from the handle and scraped the board. A sliver about the size of a bobby pin went right down the space between my thumb and thumbnail. I howled in pain to my friends, to God and most importantly to my mother. She came running from our house, pulled out the offending sliver and then cleaned and dressed the wound. After receiving the emotional equivalent of a Purple Heart medal, I went out and resumed playing.

Despite such loving care, the nail still got infected. A big ugly sore occurred overnight, and in the morning a red streak could be seen running down my hand. Mom rushed me to the hospital where, despite my valiant protests, the nail was removed. Antibiotics were administered, both topically and orally, and I was sent home with a large aching thumb, a few tears, and a trip to the local ice cream shop.

What was happening at the cellular level of this little incident bears explaining. Though Mom tried her best to clean the wound at the time of the injury, she did not get out all the pieces of dirt, debris and attached micro-organisms. None of those surviving molecules had little John's recognizable cellular uniform. It did not take long for my immunological surveillance cells to find the potential invasion; after a period of time they created the molecular equivalent of a war zone in my thumb. All kinds of cells became involved, detonating weapons so toxic that even some of my own cells were killed in the process. We call the presence of this violent battlefield 'inflammation' and we term the collective battle casualties 'pus'.

I bring this up to make a very important point. The cells in my inflamed tissues were killed in the line of duty or died from collateral damage. This kind of death is called necrosis and the flesh that suffers this process is called necrotic tissue. This death is very different from the destruction of the cells that underwent apoptosis when I was in the womb. The cells that died when I was an embryo were not killed by accident, they were killed on purpose. Thus apoptosis is more like a suicide, and necrosis is more like a wartime death. By the time I was eight years old, I was immunologically already very acquainted with at least two kinds of cell death. If we contrast these two processes with formal cell senescence (Figure 7), we understand that a description of cellular death cannot be thoroughly explained by a single definition.

This set of processes exerted over a very particular set of cells is hardly enough evidence, however, to drastically alter our overall perception of death. Is the immune system an aberration? Can we simply regard the construction of human immunity a special case and march onward with a single explanation of human death? The answer is no. As we'll see in a moment, this dying is absolutely critical to another embryonic construction project. This process, reminiscent of our immunological discussion, has to do with human identity in its most intimate form. This time, however, the death process is purely neural. And the process goes on much longer than the time period we spend in the womb.

The nervous system

The research physicians who first found this out were amazed. They had been introduced to a patient who had suffered a stroke, one of those little vascular time bombs that go off in people's heads from time to time. Certain pieces of brain tissue usually die when these things detonate. This tragedy has been a boon to research science. Sometimes such selective deaths can impair specific intellectual functions, creating deficiencies that can lead to greater understanding of brain organization. The patient in question had a whopper.

'Will you please identify anything you see in this picture?', the researcher requested of the patient, now in the laboratory, under the critical scrutiny of several neurobiologists. The scientist was holding a photograph of a rhinoceros. 'I don't know what that is,' the patient responded. The researcher held up another picture of an animal, a chicken. 'Will you please identify anything you see in this picture?' he repeated. 'I don't know what that is either,' the patient said. No matter what picture the researcher displayed, the patient was unable to identify it.

Next, the researcher changed to a different pile of cards, ones that held not pictures of animals but merely the names of animals. The first one he showed to the patient had the word r-h-i-n-o-c-e-r-o-s spelled out in

Cellular aging and death

Cells can die in many ways during the life of an organism. How their deaths relate to aging can be divided into several categories under laboratory conditions. These categories are listed below.

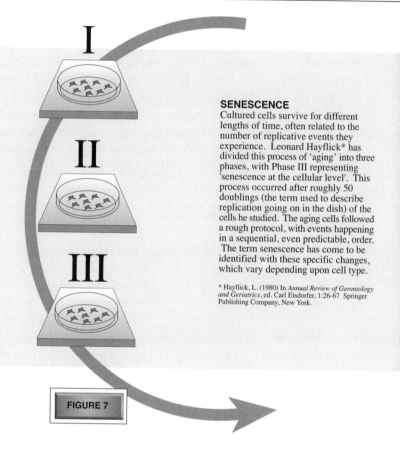

SENESCENCE

Cultured cells survive for different lengths of time, often related to the number of replicative events they experience. Leonard Hayflick* has divided this process of 'aging' into three phases, with Phase III representing 'senescence at the cellular level'. This process occurred after roughly 50 doublings (the term used to describe replication going on in the dish) of the cells he studied. The aging cells followed a rough protocol, with events happening in a sequential, even predictable, order. The term senescence has come to be identified with these specific changes, which vary depending upon cell type.

* Hayflick, L. (1980) In *Annual Review of Gerontology and Geriatrics*, ed. Carl Eisdorfer, 1:26-67 Springer Publishing Company, New York.

I

II

III

FIGURE 7

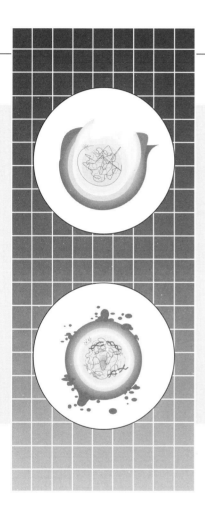

TRAUMA
This is the easiest kind of death to observe in the laboratory. The cells in the culture are wounded by an external event. This can be done chemically with certain poisons or physically (say with a razor blade). The cells spill their cytoplasmic contents, become wrinkled or simply spread out like a biological supernova.

APOPTOSIS
Apoptosis is also a form of 'aging' observable in culture dishes. It differs from senescence in that the associated aging events can be independent of the number of doublings. Moreover, the death appears to be 'programmed.' Many of the initiating genetic events in this program are well characterized (see text and also Part 3) .

large block letters. 'Will you please describe for me what this is?' the researcher asked. 'Sure,' replied the patient. 'It's a four-legged animal with a large horn on the front of its face. I think the animal is on somebody's endangered list. Something about the horn is used as some kind of drug or something.'

The scientist then handed the patient a piece of paper and a crayon. 'Will you please draw a rhinoceros?' the researcher requested. The patient then drew a crude but quite accurate picture of a rhinoceros with a really big horn. He handed it back to the researcher, who took it and waited a minute or two. Then the scientist held the picture back up so the patient could see it. 'Will you please identify anything you see in this picture?' the researcher asked.

'I don't know what that is,' the patient calmly replied.

What that was

The researchers were surprised, of course. Here was a man who could recognize certain nouns but not their corresponding graphic image – *even if he had just drawn the picture.* This deficit showed the selective and repetitive storage features of the human brain. Obviously this man had stockpiled the concept of rhinoceros in several distinct places, the text in one area, the graphic representation in another. The stroke had severed the connection between some these repositories and created his lack of pictorial recall. This deficit showed how repetitive certain kinds of memories can be, giving us a glimpse of the sophistication of normal brain structure.

Exactly how does the brain reference different areas? That's another way of asking how do the nerves know where to wire and cross-reference each other? What process does a tiny embryo employ in order to create such a magnificent structure as a brain? Believe it or not, the ability to create neural pathways is very relevant to our discussion on aging and death. The brain constructs itself in part because of programmed cell death.

Apoptosis on the brain

One of the most extraordinary links between the human immune system and neural development has to do with cellular suicide. When neural structures begin forming in the embryo, there appears to be an abundant supply of nerves. These nerves stream to many parts of the developing body, guided by forces that remain mysterious and undefined.

But what does this have to do with cell death? In many structures, there appear to be 'too many' nerves for the embryo's satisfaction. As a result, many nerve cells simply begin to die off. This dying off can be massive. In some places the carnage affects 80% of the nerves in the local area. In some places only 20% are destroyed. In still other places no nerves die. This destructive pattern occurs because of apoptosis, the same kind of programmed death found in the immune system. And it doesn't necessarily stop once the baby is born. Indeed the wiring of the human brain continues in some fashion clear into adulthood.

Why do the cells die? No one knows. In some cases it appears to be simply the way to get rid of an embarrassment of neural riches. In other cases it seems to be a way of destroying nerves that only served transient purposes. There may even be some harmful cells that bite the bullet just to prevent us from experiencing as yet undefined dangers. What *is* known is that neural apoptosis happens a lot. It is seen in animals ranging from salamanders to mice. And it is seen in the human brain.

Why bring this up?

I bring up the subject of neural development to demonstrate that this development-through-death process is not simply the domain of the immune system. When one opens up the hood of a given cell and asks 'how are you going to die?' the cell must respond 'what do you mean by die?' In this section we have studied two meanings – apoptosis and necrosis – and there may be many other cellular death protocols yet undiscovered. The concepts of aging and death appear to be as flexible and expandable in cells as they are in creatures. Which leaves, of course, almost no room for an all-encompassing definition.

The most interesting characteristic of these dying cells is that the 'aging' occurs at a time in our lives when obsolescence should be the farthest thing from our minds. Embryos are usually thought of as growth machines; indeed if we grew in our adult life at the same rate we develop in our first month of life, the average human would be more than 15 000 feet tall. The fact that cellular destruction is intimately connected to embryonic life must change the way we view terminality, straining at our ability to understand aging and death processes every time we examine a cell.

The moment of death

In these last pages, we talked about human death at the cellular level, looking internally for definitional clues. In this section, we are going to talk about death from the outside, with the same purpose in mind. To do this, I would like to begin not with biology but with history – specifically with a royal prince who seemed to have as hard a time defining death as we have.

The story starts with a blonde-haired woman, Inez de Castro of Portugal. She was the consort of Pedro, eldest son of a 14th century Portuguese monarch. The beautiful consort, a woman with whom Pedro was hopelessly obsessed, was also involved in a great deal of palace intrigue. This habit cost Inez her life; she was found murdered in her apartments, and the king ordered her buried with the fanfare always accorded royalty. Young Pedro, not yet ready to assume the throne, mourned mightily.

And, one might say, bizarrely. When the king died, an early act of Pedro's accession was to have Inez' body exhumed. It was placed upright on the royal throne of Portugal, where a crown was uneasily balanced on what remained of her head. In this royal posture, Inez de Castro was declared Queen of the land. The gentry, church elders and various subjects were served notice to come to the palace to pay her homage. They were made to kiss the bones of her hands and witness her countenance, her still yellow hair dangling from her head. One of her hands was made to grip the scepter, the other the orb of royalty. There was a coronation ceremony and at night, a royal procession, which extended many miles. Her majesty was escorted in a horse-drawn carriage to the royal abbey. There the dead queen lay in her royal robes, lit by hundreds of candles, for all her subjects to view. Eventually she was interred.

Whether Pedro was oddly political, emotionally confused or just royally weird is a matter of historical conjecture. What it illustrates for our purposes, is a bizarre alteration in the way many of us view human death from outward appearances. We might view Pedro's behavior as a psychological aberration. Looking externally, we have decided there is a point in time, perhaps a moment, when life ceased. As the decaying process so obviously follows this moment, Pedro's behavior appears 'sick' to us.

But is the idea of a moment of death a valid biological event? From a historical perspective, the moment of death is a very Western notion. And while it may be a commonly held idea, its familiarity does not make it valid. Certain new technologies, primarily developed by the West, have placed even this traditional concept on shaky grounds. To examine these ideas, we will briefly explore how several other cultures have viewed human expiration, returning to the Western technologies (and our definitions) at the end.

Antiquity

For many ancient cultures, there was not a single moment of death, but rather a moment of transition. Perhaps the archetypal example of this was found with the Egyptians, whose culture we mostly see through the lens of their funeral rites.

I am of course talking about the familiar ideas of mummification and earthly preparations for the afterlife. The belief in the early years was that most souls never left the earth, even after they died. Only the king could reach heaven, which was described as being in the company of the gods. In this period of Egyptian history, proletarian death was only a hidden form of earthly life, and the moment of expiration a temporary transition to it.

After the fall of the so-called Old Kingdom, these ideas abruptly changed. Divine access, it was said, was universal and also quite arduous. The soul of all Egyptians possessed a visa to reach the land of the gods, but entrance could be achieved only after a long, dangerous struggle. To aid the soul on its journey, it was necessary to preserve the body and also to equip it. This central idea gave rise to common mummification and elaborate burial rites. Various representations of objects needed for the journey were placed in the tombs. As a result, we know a great deal about the daily lives of the people in this amazing culture, and how, after so many centuries, we still struggle with the same mysteries of death.

The rites of ancient Greeks, a culture that derived many social characteristics from Egypt, was not nearly so elaborate. The idea that a moment of death was really a moment of transition was preserved. In many locations a sealed letter was attached to the corpse, announcing to residents of the next world that the dead had undergone the proper burial protocols. But the elaborate preparations were missing. The body was often merely washed clean with water or wine, then anointed with perfumes. There might still be a procession (with professional mourners hired to wail at the top of their lungs), but the body was never placed in something as elaborate as a pyramid. Variations on these rites persisted throughout antiquity. They were even codified into Roman Law.

In the West, the idea of a moment of transition evolved very easily into the idea of the moment of death. In the Christian tradition, the point of expiraton became a critical crossroad; the spiritual worthiness of a person's soul was assessed at the moment of death. To ensure likely passage, certain rituals were performed. For example, candles were lighted around a person's body in the belief that the deceased's soul was especially vulnerable to the assault of demons, which, as everyone knew, were also vulnerable to light. The ringing of bells at death was designed to frighten these same annoying spirits. If all went well, then the rewards for a proper earthly relationship with God would be meted out in a glorious eternity. The price of admission was experienced in the moment of death.

Sometimes a great notion

However, there were and are cultures which handled the idea very differently. Certain primitive cultures did not bury their dead the instant they 'died.' The reason is simple: they did not believe that they were dead. Even in the 20th century, many groups believe that a true departure is a gradual event, taking various lengths of time, and even requiring assistance. Some groups see this passage in terms of days. One indigenous Malaysian tribe believes this trip to take 20 years.

What different groups do with their deceased seems as strange to Western minds as the young king Pedro's behavior. It was the practice of the Dayak tribe of Borneo to keep the corpse at home, handling it as if it were alive. The body was sat up at table, given a place setting and a portion of the meal. It was offered conversation and advice, and was surrounded by the company of friends and loved ones. Only when Western society came along with their notions about sanitation was the practice stopped. In response, the Dayak built a house that their dead could inhabit until the time of burial drew near.

There are Indonesian tribes that have a similar kind of interactive waiting period with their deceased. The living relatives collect the bodily liquids produced by the natural decaying process of the deceased. These liquids are then saved for a normal meal, when they are sprinkled over rice and ingested. This ritual is thought to have a spiritual significance, where the living can partake in the death of the loved one, and vice versa.

Throughout history, there have been great ambiguities about the presence or the significance of the moment of death. When one looks externally at a 'lifeless body,' something certainly appears to be missing – no invocation of slime mold or apoptosis could tell us differently. It is perhaps natural, considering our heritage, that such an idea as the moment of death should be so firmly entrenched in our thinking. But as we'll see momentarily, it is a simple idea only as long as we don't look too closely.

Let me give two examples, taken from incidents that occurred in different countries many miles from each other.

The strange biology of Bruce Tucker

The first example took place in the United States, during a conversation between two doctors: 'Well, do we or don't we?' The two physicians looked down at the patient on the respirator. There lay Bruce Tucker, a 56-year-old laborer, silenced by a massive head injury and cradled between the sheets. Every other organ in his body worked perfectly. It was just that his brain had flat-lined (which meant no detectable brain wave). Without fully deciding whether Tucker were dead, someone placed their fingers on the respirator and turned the power switch to 'off'.

The body on the bed lay still. Five minutes later, the power was turned back on, this time to preserve the organs. Waiting transplant surgeons were called and within the day, Bruce's heart was beating in someone else's body. His kidneys, too, had been removed. The business of organ transplantation, though young at the time of Tucker's death, was already beginning to save lives that would otherwise be lost. As Bruce's final act on earth, such a donation was a noble and fitting thing to do.

As long as everyone was sure that Bruce was dead.

The trial

The problem was, not everyone was sure. His brother was not, and put teeth into his objections with a lawsuit. He accused the transplant team of a 'systematic and nefarious scheme to use Bruce Tucker's heart and hasten his death by shutting off the mechanical means of support.' Even though a neurologist could not find any brain waves, the case went to court. To the angry disappointment of the living Tucker, the jury found in favor of the doctors. A member of the transplant team remarked that the decision brought 'the law up to date with what medicine has known all along, that the only death is brain death.'

Unfortunately, the time that has passed since then has shown that we still have ambiguity. If it were all that simple, you wouldn't be reading this chapter.

The provocation

Of all the moral issues forced into the 20th century by medical technology, life and death issues are perhaps the most painful. The Bruce Tucker case was a landmark decision, setting a precedent about the role of the human brain in the final definition of human death. From a biological point of view, however, the case has solved nothing.

The problem is that humans are multi-cellular creatures too, just like termites and trychonymphs. Some groups of human cellular systems continue to function normally even when other groups fail. The heart will insist on beating even if the brain is dead; heart tissue can even find its rhythms if it is removed from the body. As mentioned previously, we don't age and die because all of our cells throw up their collective hands and say 'I quit!' Since human beings are composed of individual cells, not all of them possessing the ability to age us into oblivion, which cells are we going to use for our definitions? When it came to the Tucker case, the courts decided that the collective action of brain cells was good enough. And with that decision, the cherished Western idea of a moment of death became permanently attached to a neural substrate. Not that it was a catch-all definition, as illustrated by a resuscitative technology that was emerging at the same time as the decision.

It all had to do with our second example, which concerns a young boy in Norway. To describe his impact on the 'moment of death' idea, his story is related below.

Of all the nerve

It was a bright morning, one of those incandescent days when the sun shines on the snow and makes one think of laundry detergent commercials. The young adolescent Norwegian boy decided to put on his ice-skates and spend the day sliding around a frozen river. As the day wore on, the more fragile ice began to melt, becoming dangerously thin in certain places. As afternoon turned into evening, the skaters were warned not to skate too closely around those places, lest they fall through a crack.

But the young boy did go too far. The ice cracked and he fell into the river, hitting his head on the ice as he went down. There he lay, submerged in the freezing river. It was more than 20 minutes before someone noticed. Because of the delay, it seemed too late for the medical authorities who eventually arrived; the boy had no pulse, no breath, no outward signs of life. And this might have been the end of the story, one more casualty of a typical Norwegian winter, except for the presence of two biological miracles.

The skater was rushed to the hospital where drastic emergency procedures were initiated. Heart massages, drug injections, blood transfusions, the entire technological battalion was sent to the rescue. To the doctor's complete surprise, the first miracle occurred – at least in part. Some of the boy's vital functions spontaneously sprang to life. A heart beat suddenly leaped onto the monitoring devices. And the boy began breathing on his own. With this turn of events, it seemed that all might be well, that perhaps many functions were salvageable and the boy would live.

It turned out to be a temporary miracle, with only some of the biology permanently resurrected. The skater soon lapsed into an unconsciousness so deep that his brain flat-lined again, no detectable brain waves on the EEG. In the sterile words of a clinician's evaluation, the boy had become decerebrate.

Even so, the medical authorities kept him on life support, because he moved in and out of this waveless state. It wasn't until the fifth week of this awful equilibrium that the second 'miracle' took place. For reasons that are still unexplained, the boy's brain all of a sudden switched back on. Electrical activity began to be detected, even slight but real motor activity was observed. And these functions stayed active, with a vigor indicating that there would be no relapse. Very shortly he awoke and began to grow stronger by the hour. And by the day. Recovery was so quick that in six months' time, the Norwegian skater was back at home, with very little outward signs of his trauma.

In the parlance of the clinician, the skater had survived what appeared to be irreversible brain loss. In the words of his pastor, the boy had come back from the dead.

The definitions

This incident, like so many since, brings to focus a question asked ever since brain waves and expiration were paired: how valid is the linkage? This is a question we are about to address. For our purposes, it is another example of the problem in finding an overarching definition of the ending of human life.

The issue seemed settled as far back as 1968, when a group of Harvard physicians codified the criteria for a legal-friendly definition of human death. Their model could be divided roughly into three parts.

(1) The patient must display a flat EEG for 24 hours;
(2) After a lapse of time, the EEG must be checked again;
(3) If both readouts show the same lack of activity, the person can be pronounced dead.

If it could be determined that upon resuscitation a 'brain dead' patient would only lead a vegetative existence, the person would possess an irreversible coma. Such a patient would be said to have expired, too.

As can be predicted from the Norwegian example, this definition has some problems. But the validity of these criteria was challenged not only from observing revivable patients, but also from new technology. Several years after the standard was issued, a neurobiologist working in California developed a device that was several orders of magnitude more sensitive than the contemporary EEG machines. Most of the then-current devices measured activity only at the surface of the scalp. This new machine could measure activity deep within the folds and twists of the brain.

The field tests with this device showed some astonishing findings. In one examination, more than 26 people were reclassified as living when the new machine was applied to their skulls. A number of these individuals were actually revived, and, like the Norwegian skater, showed very few signs of brain damage. This of course led to the question: how many individuals were lost through use of the old machines? Most amazing was the flexibility of the death definition, and how easily it could change with the simple application of new technology.

This did nothing to smooth the controversy about utilizing neural substrates to define the moment of death. And in the mid-1970s, the sensitivity of the new technology was pushed to an almost absurd limit, making the situation worse. Under the watchful eye of a real live neurobiologist, this actually happened:

The scientist bought some lime-flavored jello. Using a mold roughly the size of a human skull, the researcher created a 'jello brain' and then brought it to the hospital for study. Using the new sensitive EEG machines, he hooked up the jello in a standard hospital intensive care unit and then applied EEG analysis. To his delight and to other people's horror, the clinician actually detected electrical activity!

The doctor was not making scientific history, but neither was he making up his findings. He was able to demonstrate that the machine was actually detecting stray activity from the various electrical machines in the room – it even responded to the attendants entering and leaving the unit. The researcher made his point very clear: this electrical definition of life is very fragile when applied to nerves in the brain, subject even to background noise. To make an accurate reading and not identify something as alive when it is actually dead, one must be able to filter out such noise.

The problem is that the technology, and indeed, our knowledge of neural activty, isn't all that discriminating. If you eliminate all the noise, how can you be sure that you are not also eliminating some crucial flicker of activity – an activity that might lead to the full recovery of a loved one?

As of this writing, the question of when the Western moment of death occurs has not been solved. It has led to an interesting ambiguity and something of a legal nightmare. Exactly when someone is pronounced dead truly depends on the place he or she died. In the US, some states say that the brain activity is the sole requirement. Other states say that respiratory and cardiac activity can be used in place of the brain requirement. Still others ignore the brain requirement altogether. In France, the brain has to be silent for 48 hours. In the days of the former Soviet Union, Russian patients only needed to be flat-lined for five minutes. In the words of Dr Henry Beacher, the clinician who headed the original Harvard study group: 'Whatever level (of electrical brain activity) we choose, it is an arbitrary decision.'

And with that as an epitaph, so may be an overarching definition of the moment of human death.

So then what do we do?

If the moment of death is so hard to define, does that mean it is impossible to define? Surely the process takes place. There are grave markers, embalming rituals and funeral services. The process is so inevitable that humans can actually prepare for it. Does death have to be one of those things that nobody can define but everybody can easily discern when it exists?

In this section, we have attempted to debunk an idea that seems, at first glance, intuitively attractive: the presence of a far-reaching, common-sense definition of human aging's primary goal – absolute death. Experts have not left us with a defined answer to the problem of the moment of death; they've only left us with a question. This is an idea absolutely critical to the apprehension of the rest of this book. Human or otherwise, death has no central absolute definition. The monolith, for better or worse, falls with a thud.

Conclusions

Even with its absolute relevance to a discussion about aging, we need not be uncomfortable with death's ambiguity. We can be very familiar with the reality of events we cannot define. We know that as we age, a very profound process is playing out in our bodies, we just have no adequate way to categorize it. Indeed the very attempt can change the nature of the moment. Such nearly undefinable events happen in art all the time. Sometimes they even happen in science, especially when we try to bring antiquated definitions to new discoveries.

What do we conclude about death, human or otherwise? Human death is an example of a very definite phenomenon that has a very indefinite description. Because a critical part of an aging definition hinges on our ability to define death, we have sought to explain the process in concrete terms. Though the definition remains crucial, it unfortunately also remains elusive. This is probably the result of an artificial human attempt at organization. Not that nature cares. Giraffes, palm trees, slime molds and your relatives will continue to live out their life cycles regardless of our definitions. We may just have to content ourselves with a consensus approach. In this book, we will not necessarily know what death is, and thus fully what aging is. We may just be able to identify it when we see it.

Perhaps such tasks are best kept with the poets anyway. Keats would understand its definition, even if only by experience, his immune system useless, his nerves on edge, his medical treatment archaic and dangerous. In the end he may have characterized this part of the definition better than any scientist.

> My spirit is too weak – mortality
> Weighs heavily on me like unwilling sleep,
> And each imagin'd pinnacle and steep
> Of godlike hardship, tells me I must die.

3

Why age at all?

There are many stories surrounding the life and death of Billy the Kid. Most of the stories are false. The media, no less hungry for a tabloid story in the 19th century as in the 20th, freely interspersed a few grains of truth with Wild West superstition. About the only place where the legends and facts agree is this outlaw's age of death – 21 years, 7 months and 22 days. The larger-than-life stories sadly obscure the real history of this man's short life. It is much more poignant than heroic, more melancholy than malevolent.

Billy the Kid's real name was Henry McCarty. He wasn't born in the Wild West, he was born in a Manhattan tenement slum. To escape poverty, his Dad moved the family to Coffeyville, Kansas and very quickly died. Henry's mother, now a single parent, had the daunting task of making a living in the gritty cruelty of the mid-19th century American West. It was a testimony to her resourcefulness and intelligence (and the genesis of young Henry's undying devotion) that she was able to make a go of it. She supported herself and her son in business – operating hotels, laundries and boarding houses, and even dealing in real estate. By all accounts Henry grew up well-loved and increasingly well-supported. If it hadn't been for the visit of the now-familiar *Mycobacterium tuberculosis* on the McCarty household, there might never have been a substrate upon which to build this legend.

Tuberculosis struck Henry's Mom in 1871, when he was only 12. Like John Keats in another world and another decade, she would die by slowly coughing her lungs out of her chest. It took two years for the death to occur, Henry eventually dropping out of school, trying to shoulder some of the practical chores of their lives. In the last several months of her life he did not leave her side, and when she died, Henry went out and got drunk.

And stayed that way. In fact, he became an alcoholic. It wasn't long before the young man's life began to unravel. Losing much of his assets in gambling, he began to travel with a rough crowd, which in those days harbored some fairly mean characters. They were not kind to young Henry. Soon penniless, he was often derided or made fun of because of his youthful appearance. It was at this point that he changed his name to Billy Bonney, which quickly became Billy the Kid. The first sheriff to arrest him said Henry changed his name so his mother's reputation, highly esteemed at the time of her death, would not be soiled in memory.

At age 18, Billy the Kid shot his first man, a blacksmith who humiliatingly slapped him around the local saloon. He immediately gained the reputation as a teenaged killer. Frightened of this action, he ran out of town and headed toward the New Mexico–Arizona border. There he drifted in and out of the various cow camps, eventually learning the trade of cattle rustling. He drank more, gambled more and lived the life of a cattle-stealing vagrant. And that might have been the end of this legend, no movie script required, if it were not for a chance encounter with an educated English gentleman rancher. After that, the clock of his life would begin to wind down.

The rancher's name was Tunstall. He was a humane gentleman who took an instant liking to the boy. He hired Billy to do odd jobs around the ranch. Billy warmed to the man's praise and affirmation. He began to idolize the rancher, and work hard to please him. The relationship became so positive that Tunstall remarked 'That's the finest lad I ever met. He's a revelation to me every day and would do anything on earth to please me. I'm going to make a man of that boy yet.'

It was not to happen. Ironically, Billy was to gain the last part of his reputation defending Tunstall against those in his former profession – cattle rustlers. A gang of outlaws demanded Tunstall surrender half his herd. When the unarmed Tunstall refused, they shot and killed him. Something in Billy broke that day in the late 1870s, and he swore a blood oath to kill everyone involved in Tunstall's murder. This included Jim Brady, the local sheriff who had been bribed by the outlaws not to interfere.

And kill them Billy did. The slaughter led to a massive manhunt in which Billy was found and arrested, ironically by a former friend, a cattle-rustler-turned sheriff named Pat Garrett. Sheriff Garrett hadn't reformed as much as he had grown clever, exploiting both sides of the law for personal profit. He found The Kid in a ranch-house, and shot him through the heart.

The myths began almost immediately. Within three weeks a book was out about the life of Billy the Kid, exaggerating the number of people he killed and amplifying his legend. Billy's body went on display in a traveling road show for a paying public. The sheriff himself eventually collected a reward (though he was a salaried lawman) and wrote his own book, further distorting the material facts of the arrest. The eventual mythology of Henry McCarty's short life percolated into the 20th century with more books and even a couple of movies. None chose to tell the real story.

An important question

The reason for describing the death of this young man is a familiar one. As with the movie star Rudolph Valentino and the poet John Keats, there are no memories of an elderly Billy the Kid, no photographs of a wizened outlaw talking to his grandchildren about a checkered past. What makes this story saddest is that, barring a TB infection, the Kid's early demise might have been avoided. Billy's death was a consequence not of a burst appendix or bacterial infection, but rather of an already violent environment. It is upon such violent environments that this last chapter will focus.

Though we like to think otherwise, such everyday violence is not just an aberration of American history, it is a general fact of natural history – and with the same, youth-selecting results. An African lion will stalk a young wildebeest with all the eager cunning of a Western sheriff. Wasps will tranquilize spiders before laying their eggs in them with the express purpose of allowing their young to dine on a living lunch once born. Male elk will wound and even kill each other in battle for the right to mate with the females in a given herd. With all this natural violence, it is no wonder that we seldom see aging creatures able to survive.

This fact leads us to a very important question. If young and healthy organisms are the ones that have the best shot at survival, given millions of years of evolution, why does aging occur at all? Since we are so familiar with the aging process, we might take its inexorability for granted so that we do not question its mechanisms or ponder its purposes. And yet, as we have learned, other creatures have the capacity for immortalization. So do human cancer cells. Given that youth has the best chance at survival, why haven't all creatures jumped on the eternity bandwagon?

These are questions we will try to answer in this section.

A framework

The purpose of the last three chapters has been to place the process of aging and death into a context that makes sense. We have spent some effort attempting to provide a definitional framework for the phenomenon of growing old. In this last section, we will attempt to provide an evolutionary one. We will start with a general review of the basics of natural selection, discussing the aforementioned cruelty with an eye on evolution's grand prize – survival. Then we will take a look at how other researchers in other decades have attempted to answer the question about the reason for aging. We will begin with a personal friend of Darwin's and end with an interesting idea about cancer.

All told, this discussion will be our final comment on context before we start talking about the hows of The Clock of Ages. Understanding the selective pressures of the natural world will place into perspective the research directions undertaken by scientists from around the world.

The hows of natural selection

In this section I will be using the words 'selective pressure' a lot. I would like to explain its meaning by giving a personal example.

I live in the city of Seattle. Like most American urban centers, it has an inner city filled with both tragedy and promise. It also has 'rich' areas filled with tax shelters and BMWs. The amount of money it takes to live in the Inner City is very low compared to the amount it takes to live in the upscale areas. This disparity lies near the heart of many of America's deepest social problems.

If you want to live in Seattle's rich environs, you're going to need a certain amount of income. If you don't have the income, you won't be able to afford the housing and taxes. If you had the income and then lost it, you would probably have to move out once your savings were gone. There is a selection that is occurring in those places, a pressure as it were, exerted on the resident population to produce or sustain a certain level of income. Selective pressure is what maintains the characteristics of the local populace and if you want to survive there, you'd better have the necessary attributes. Which in this case means money.

Most environments in nature possess just as real survival pressures within them as do the rich areas of Seattle. Only these pressures don't involve income, but rather biological characteristics. Take for example a forest where all the leaves are very high on the trees. Let's say you are a land-bound, non-tree-climbing mammal who requires those leaves in order to live. If you want to live in that forest, you'd better have a long neck. If you don't have a long neck, you are either going to have to depart the forest or die. The forest has exerted a pressure on you, a selective pressure to possess a certain set of biological characteristics every bit as powerful as an annual income.

This selective pressure can be long-lasting, propelling organisms into a genetic future or a selective dead end. If you have a long neck, you will survive, which means at some point you can start reproducing yourself. Your colleagues with short necks will not have this luxury in the forest; they will have either moved or died. If the genetic roll of the dice gives your progeny long necks, they will be on the same happy path you have traveled. If you produce children with short necks, they will meet the same fate as your soon-to-be-extinct colleagues. This selective pressure is embedded within every biological environment. It is just as rewarding and just as arbitrary as any housing situation in any city.

Selectivity is the engine that fuels the ideas of Darwinian evolution. It is welded into the fabric of the world of biology and the only process that gives observational biology a context. Stated simply, it says: The species best suited, or adapted, to its environment is the most likely to survive (for a more detailed explanation, see Figure 8).

This process of selection is extremely effective. Powerful organisms become locked in an uneasy struggle with their environments, with gene selection occurring over millions of years within a specific location. Some environments change very quickly – a violent upheaval in an earthquake, a sudden and permanent flooding, a volcanic eruption. Such events can produce catastrophes for organisms evolved to live in a place that had previously been stable – and their survival track record bears silent witness. It has been estimated that 99% of the organisms that have ever shown up on the planet are now extinct.

An intersection

In the middle of this delicate balancing act, we must contend with the process of aging. How do environmental pressures interact with organisms

A quick review of

Natural selection

Darwinian evolution is powered by the theory of natural selection. The theory can be divided into four major parts, as listed below.

REPRODUCTION
Species reproduce. They create members of their own kind, with their characteristics generated by their genetic codes.

VARIATION
Reproduction occasionally introduces variations in the species We call such variations mutations, which can be directly attributed to changes within the creature's genetic code.

ENVIRONMENT
There are limited environmental resources available to any given creature. These constrained resources include food supply, water, living space, etc.

COMPETITION
Each individual must compete for the limited environmental resources. Those creatures with the genetic characteristics most advantageous to extracting nutrients and energy from their environment will survive.

FIGURE 8

to produce their 'natural' longevity? Or do they? Almost every other biological characteristic embedded into creatures is subject to selective
demands, from leg length to the ability to see certain colors. Is there any
reason to assume the aging process has escaped such surly pressure?

Many great minds have attempted to answer this question. From the
dawning of Darwin's ideas to the present-day practitioners of molecular
biology, researchers have postulated, hypothesized and argued about
aging's overall role in life. The question remains mostly unsettled. In this
section, we are going to summarize certain researchers' thoughts, beginning with theories about environmental stresses brought about by the
interactions between parents and progeny. First to be considered will be
the ideas of Russel Wallace, a 19th century Englishman who is also the
largely unsung co-discoverer of natural selection.

The ideas of Wallace

Russel Wallace was very interested in the interactions between parents
and children, specifically in terms of competitive survival. He encapsu-
lated his theories with a thought problem – imagining an immortal
creature, living in a stable environment, which did not reproduce.
Wallace naturally reasoned that any death would be accidental and,
with an absence of progeny, signal the end of the species. He therefore
concluded that natural selection favored reproduction over agelessness.
As wonderful as immortality would be for the individual, progeny
would be more valuable for the group.

Then why not have an immortal organism that reproduces? Or maybe
not immortal, but simply long-lived? Wallace once again used a thought
process to address these questions, asking what would happen if groups
of parents lived a long time in a specific environment. His projection was
simple: they would start to compete for that all-important and extremely
limited food supply. The parents would thus become a threat to their
children by virtue of their longevity, imposing their own needs on very
finite fuels. Everyone would soon starve to death, or die of thirst or pollute
themselves with their waste, whatever. Immortality, or even a very long
life, would be selected against in the natural world. In Wallace's own
words:

Natural selection therefore weeds them out, and in many cases favours such races as die almost immediately after they have left successors.

(quoted in Weismann, 1889, p.2)

Wallace's ideas squarely describe aging as a process subject to the pressures of natural selection. They've even been interpreted as a collection of theories loosely entitled 'group selection' (which they're not), notions we will discuss shortly. Because of his confusion about individual vs group dynamics, Wallace has sometimes been severely criticized. Nonetheless, he introduced the idea that individual immortality could be sacrificed for increased survivability. Aging would then come into existence for a simple reason: competition between parents who may live a long time and the progeny which must survive is an altogether unhealthy thing.

Weismann

Other scientists after Wallace attempted initially to refute and ultimately to extend his arguments. One of the most influential theorists in the explanation of evolutionary biology and aging was August Weismann. He was the originator and early proponent of the idea mentioned previously, the theory of group selection. He dropped the notion altogether in his later life, and came back more to the explanations first proposed by Wallace. He even anticipated that a mathematical analysis of selective forces with respect to aging could be described. More than a hundred years later, this idea has taken hold.

An example in army ants

I would like to give an example of this group selection idea by way of an insect analogy and a fictional story. I once read a tale about the life cycle of army ants and a wheel-chair-bound man.

The story revolved around a paraplegic scientist working in the jungles of Africa. He was near an active volcano, which eventually erupted and sent the research station into a panic. The scientist's colleagues hastily bundled him, still seated in his wheelchair, into the back of the truck. They sped down the trail as fast as possible, and promptly smashed into the side of a tree. The truck overturned near the bank of a river and all the members of the team were killed. Except the scientist, of course. He was trapped underneath, his arms broken, his body still tethered to his wheelchair. The story involved the struggle between the scientist, the threatening volcano and the stars of the story, the army ants, who were conveniently marching along the same trail.

The behavior of army ants is an interesting metaphor for Weismann's idea of group-over-individual selection. In the life cycle of this cooperative insect, personal welfare has long since been sacrificed for the good of the group. Army ants march in determined columns, they attack as a group, they protect and reproduce their offspring in a community effort. Even when they settle down for the night, they do not build nests. Rather, individuals will grab each other by their jaws, forming complex foot-to-mouth structures that actually *become* the shelter for the rest of the group. So cooperative is a team of army ants that some biologists don't describe them as a large collection of individuals. Rather, they are viewed as a complex single organism made of many parts. As can be seen, the selective pressure being exerted on any one individual is not nearly as important as the pressures that affect the collective. Which was Weismann's idea. The power of these forces is observed simply by observing the destructive capabilities of a group of army ants.

Or the wide-eyed terror of a little boy reading about a trapped scientist fighting for his individual life in the depths of central Africa. He was successfully rescued, by the way, a fact that for years my nightmares completely ignored.

But back to Weismann

Weismann's initial idea, as with so many important works, began with a rejection of the current dogma. He refused to believe that the life span of a given organism was solely determined by its individual physical and chemical characteristics. He wrote:

> Physiological considerations alone cannot determine the duration of life. (The) duration of life is really dependent upon adaptation to external conditions, that is length, whether longer or short, is governed by the needs of the species.
>
> *(Weismann, 1889, p. 10)*

Weismann attempted to illustrate his ideas by examining the life cycles of many insects, and also mammals and birds. Aging, in Weismann's mind, had to be assessed only by examining the worth and survivability of the whole, ideas which also found home in certain contemporary political theories. To the collective sighs of the elder gentry of his day, Weismann did not consider an older individual harmful. He just thought that individual immortality would give no corresponding advantage. He also felt that it was a somewhat impossible task anyway, given the rather delicate nature of terrestrial creatures.

This idea of the evolutionary deterioration of a neutral characteristic was revolutionary. It was also somewhat in contradiction of Wallace, who placed great emphasis on the problems immortality might hold. Yet the connection of this neutral deterioration to the problem of aging – a process Weismann called panmixia – was to have a lot of power. In fact, it made Weismann change his mind.

A prophecy in numbers

What diverted Weismann's thinking was the notion that the effort necessary to keep an organism alive is only worthwhile as long as the creature is reproductively active. He even began to see a possible detriment an elder organism might confer on the whole. At one point he suggested that humans might be able to calculate just how natural selection could affect a creature's life span. What was needed was

> the statistics of destruction, that is the probabilities in favour of the accidental death of a single individual at any given time.
>
> *(Weismann, 1889, p. 156)*

This sentence suggested a mathematical approach to the evolutionary problem of aging, one that might give predictive power to the normal thicket of hypothesis and postulate so prevalent in his day. Weismann's approach was to be adopted by others, using much more sophisticated technology, almost a century after he proposed it. We will explain some of these ideas in the section following.

The mathematics of the aging process

The tool that Weismann needed but did not have was the mathematical framework of population genetics. It really didn't come into fashion until several years after his death. Had he known about the substance of this then-new discipline, his ideas might have evolved even further than Wallace's.

But others did extend Weismann's ideas in mathematical directions. It is beyond the scope of this book even briefly to summarize the mechanics of population biology, however. Suffice it to say that population biologists are interested in predicting the behaviors and fates of whole groups of organisms. They ask questions like 'how many organisms will be left after a period of time?' or 'what is the reproductive value of creatures when they reach a certain age?' The discipline has always been concerned with senescence issues.

Before we leave the subject of population biology and aging, we shall mention one more theorist, a researcher deeply interested in the role of aging in survivability. His name was Sir Peter Medawar, and he put into words and numbers the central purpose of aging in evolution.

Medawar

Medawar described in arithmetical and graphical terms an idea that has been obvious to just about everybody since the biblical Abraham and Sarah. As you get older, the reproductive contribution you make to the ancestry of the future decreases. Basically, Medawar was saying the same cruel things Weismann was describing in his last years. If some disaster strikes a person who has passed the age of reproductive fitness, so what? The consequences are by and large unimportant to the survival of the rest of the species. Thus, and here comes the important idea: *the force of natural selection declines with age.* It would even be true in a theoretically immortal population, provided we were realistic enough to consider the true hazards of mortality.

The idea that the aging process is untethered from the normal pressures of survival has been embraced by many contemporary evolutionary biologists. It has an intuitive attraction. Since death by disease, accidents and normal predation usually happens long before an organism gets old, there would be no reason to *have* an aging process. As long as a proper balance between opposing biological forces was maintained in a particular environment, the population would be controlled by interaction, not by senescence.

The idea that natural selection doesn't care one whit about aging has most profound consequences for its evolutionary context. And for the genes and the characteristics those genes may confer on an older organism. In essence, the rules change. For example, a young organism could have a gene that would do two things:

(1) Enhance its ability to survive while reproductively fit; and
(2) Kill it off quickly as soon as it was no longer reproductively fit.

If natural selection doesn't operate when we get old, there is nothing to stop such processes from 'switching on' once we reach maturity. Of course, there is also nothing to stop it if the opposite happens. The gene might confer upon the organism some really nice trait in its old age too. The point is that natural selection doesn't care. This means that aging is an evolutionary landfill of potentially confused genetic processes, subject as much to whim as to direction.

Since we are one of the few organisms on the planet who regularly live longer than we should, such whims are excruciatingly relevant.

Still more lessons from a tumor

This idea of an evolutionary landfill has recently found an interesting slant. It all came from an idea regarding the now familiar connections between aging and the biology of cancer. I will describe its role in the 'reasons for human aging debate' next.

There are those who say that this untethering of selective, regulatory processes isn't quite true; that aging is a consequence of an accident, but not due to randomly activated genes. This proposal deals with discerning a purpose by examining survivability not of whole organisms, but of individual cells.

As mentioned previously, tumors form because some event jerks the reproductive cycle of a normal cell into molecular overdrive. The cell will start to divide, and then, instead of resting, will divide again. And again. If the pre-tumor can coax blood vessels to come and feed it, a tumor will form.

There are genes which, if mutated, can give a person cancer. A number of these genes have been cloned, and we learn more about the life cycle of a cell with every gene we isolate. By determining their normal unmutated functions, we learn more about the genetic processes supervising the life cycle of a cell. If the mutation is present, the individual may be at risk for certain kinds of tumors; if the gene is absent, a low risk may be presented.

It is of course not the function of these genes normally to give someone cancer. Their job description is to monitor the cell cycle, making sure just the right amounts of molecules and pathways are in place and functioning. These regulating genetic sequences fall into two general classes: those that stimulate reproductive cycles and those that suppress them.

What does the presence or absence of these tumors have to do with the cell cyle? As we will see in later chapters, cells grow old and die just like people grow old and die. And the older a human gets, the more susceptible the person becomes to specific kinds of cancer. The question of why those events occur is as valid at the cellular level as it is at the creature level. There are researchers who believe they have an answer that fits nicely into an evolutionary context.

Taking care of business

The answer has to do with how individual components react when large processes are foisted upon them. What that means can best be illustrated by an analogy drawn from the business world.

As companies have gone multi-national, they have attempted to introduce their name-brands to people who not only speak different languages, but also have different idioms. This has sometimes proved disastrous for the business. For example, the Chevrolet Division of the General Motors car company introduced their best-selling car the Nova, to Spanish-speaking Central America. They had no idea that Nova, which could be read, No va, meant 'won't go' or 'it won't go.' Red-faced, the company quickly changed the car's name to something more marketable The Johnson Company started selling their furniture polishing product 'Pledge' to the Netherlands. Unfortunately, in Dutch the name means 'urine,' making it extremely difficult for the general populace to ask for it by name. The most famous cultural *faux pas* probably belongs to the Coca cola company. They expanded into the Chinese market in the 1920s. In trying to keep a consistent name, the officials chose Chinese characters that, when spoken, would sound like the name Coca cola. Unfortunately, they did not *mean* the same thing. The message their chosen characters announced to the world was 'Bite the wax tadpole' or 'wax-flattened mare.'

The company soon changed the characters. The drink now means 'good mouth, good pleasure.'

The hypothesis

These advertising goofs were embarrassing (not to mention costly) because a single large mistake was introduced to many individuals. This interaction illustrates very well what some scientists believe about the aging process and cells. They believe that a 'mistaken' process is let loose on cells as they age, and the result is as disastrous as a bad advertising campaign. What is this process? It is cancer-protection, useful in a younger year, now gone awry.

These scientists say that aging occurs to protect complex organisms from a cancerous premature death. Or at least slow down a march which might otherwise lead a reproductively fit individual to the oncology clinic. Why? It has been obvious for years that the aging a cell experiences in culture irreversibly limits its proliferative tendencies; senescence is like giving the cells a permanent molecular hand brake. Because aging prevents their cell cycles from going haywire, the risk of cancer in these cells is minimized. Projected over eons of time, this hypothesis identifies aging as a generalized protective mechanism against cancer, working at the level of the cell. And in favor of younger people. Indeed, most of the experimental evidence in support of this idea comes from cell culture experiments. There is an interesting symmetry to this hypothesis: the reason why we age is to prevent us from dying by other means.

Cancer and reproductive fitness

This idea of neoplastic protection has consequences for the overall survivability of a species. If the hypothesis concerning aging and cancer protection is true, we have a 'reason' why the aging process exists. But does assigning a 'reason' contradict the notion of aging's irrelevance to natural selection? The answer, of course, is no. The aging process may just be an accident experienced by chronologically older creatures – the by-product of a process designed to protect younger organisms from cancer. Once a creature is no longer reproductively fit, the selective pressure is lifted. But the anti-cancer device ticking in its cells is still quite operative –

and there is nothing to protect the creature against its secondary effects as it gets older. We referred to the aging process earlier as an evolutionary landfill. Our deterioration may just be the result of genetic toxic waste, of great benefit to ourselves only when we are young.

There are of course many problems to be reconciled in this hypothesis. Can *all* of the deterioration seen in aging be explained by cell cycle determinants that have lost their primary utility? What causes the reproductive capacity ultimately to fail in the first place? Why does the incidence of cancer *increase* in older people? These and other questions await further insight and experimentation. The greatest contribution of such work has nothing to do with the particulars, however, or even its overall idea. This hypothesis places evolutionary processes squarely on the shoulders of events happening inside the cell. This is unlike most of the explanations we have been examining, which tend to emphasize events that occur to whole organisms. Narrowing the focus actually expands the idea, tucking internal cellular aging mechanisms into their very own historical context. It is the first of undoubtedly many hypotheses to give a uniquely cellular, rather than just an organismal perspective. How and when these ideas coalesce to reveal a single point of view will be one of the most exciting events in all of science.

In summary

This entire chapter has functioned like an arrest warrant, serving notice that our cherished definitions about aging and death may need to stand before a scientific judge. The discussion started innocently enough. We mentioned earlier that we might get a clearer handle on aging's intent if we could understand death, its ultimate end, in a clearer context. We discovered in the first two sections that the definition of death, invertebrate or human, was not going to afford us such luxury.

With the same purpose in mind, we attempted in this chapter to place aging and death in an evolutionary context. We examined the roots of natural selection and then discussed aging from a historical and cellular perspective. Surprisingly, we discovered there may not be a central biological purpose for aging. Indeed, our deterioration may simply be the evolutionary burp of genetic processes we normally use to shield our reproductive potential from harm. The scientific judge may be pronouncing a death sentence on our prejudices, with no offer of restitution for what we have left.

If it is a disappointment, it is only because we have grown up believing many myths, rustling them into our biases with enthusiasm. The reality of aging may be very different, the ideas harder to define, the purpose less easy to see.

It is a biological luxury, perhaps, to discuss such abstraction with a degree of health. We can contemplate these ideas only because we have learned to curtail – through medicine, public health and common sense – the processes that would normally destroy us. The very fact that we can understand such ideas is because we are evolutionary outlaws, having skirted the normal forces of natural selection. Fortunately, the earth hasn't destroyed our species, like some wily Western sheriff, in the middle of our reproductive prime.

PART TWO

How do we age?

INTRODUCTION

In the world of strange and unusual wills, the last requests of Charles Millar must rank as the most capricious of all time.

In life, Millar was a quiet man. Friends even described him as strait-laced. But he had another, far more interesting side. Millar was a Canadian attorney and, unknown to many, fabulously wealthy. Probably reasoning that death made these quiet constraints pointless, he funneled all his capriciousness, humor and satire into his last will and testament. This was accomplished by donating large sums of money with interesting conditions attached. In so doing, Millar showed to the world its desire for greed, and the great lengths to which people would go, to receive money – even from dead people.

Some of his victims were professional acquaintances. He had friends, for example, who were judges and preachers. Several of them had a vocal and quite public opposition to organized gambling. One judge always ruled severely when cases involving gambling came through his court. One preacher gave fiery sermons on the spiritual degradation inflicted by the addictive demon. To these two individuals, Millar gave very profitable shares in a racetrack. But there was a hitch. If the men accepted the shares, they would automatically become members of a horse racing club. Unflinchingly, and without regard to their moral compasses, the men accepted the shares.

There was another group of ministers who preached not so much on the sins of gambling, as on the sins of alcohol. They were of course tee-totalers, some even going so far as to walk on the other side of the street to avoid a bar. To this group of ministers, Millar bequeathed highly lucrative shares in a local brewery. There was only one teetotaling minister who refused the gift.

The most famous line of Millar's will, however, doesn't concern a vice. It concerns infants. The press, when they found out about this section of the will, dubbed it 'The Baby Derby'. Millar stipulated that a large chunk of his fortune should go to the Toronto woman who '. . . has given birth to the greatest number of children at the expiration of ten years from my death.'

And they all had to have the same father.

The guardians of the social forces of Canada, not to mention Millar's relatives, were outraged and the line in the will was bitterly contested. But Millar was crafty, and had used his legal expertise to make the requests in his estate watertight. The courts repeatedly upheld the validity of the will and the derby was on.

Precisely ten years after Millar died, Toronto's Surrogate Court awarded half a million dollars to four fertile women, each of whom had nine children. Though that might seem enough pranks to fill several life times, Millar had one last joke to play. The last stipulation of the will required the women, prior to receipt of the money, to publicly promise to use birth control.

And promise they did. They got the money. And Millar got the last laugh.

This part

Age has a funny way of exposing interesting motivations, and our death wishes can be just as revealing. Millar's will can almost be described as performance art, a biting satire into the power of greed over principle. His joke got a lot of publicity, and in so doing, seemed to stretch his life beyond the grave. It gave him a form of immortality, at least if you read history books about famous wills.

This desire to live in some form beyond the grave is common. In an effort to short-circuit the aging process, many people have desired to leave some permanent monument that outlasts their biology. In this section, we will explore the aging and death of eight famous people – artists and humanitarians who have achieved a form of immortality by leaving an indelible mark on their century and our world. But only a form of immortality. Even within their genius they discovered the same thing we do everytime we look in the mirror: as physical characteristics change with age, our mortality springs on us suddenly, like a bad joke.

But exactly what kinds of biological alterations exert such a profound change in attitude? To what does the will to survive (whatever that means) react? What happens to our bodies as the years go by? We will explore the nuts and bolts of The Clock of Ages in an attempt to answer these questions. That is, we shall discuss what happens to the various tissues in our bodies as we grow older. Starting with skin and hair, we will examine what the years do to our outward appearance. Then we will move inward in an increasingly intimate fashion. Muscles and bones and joints will be examined. So will the heart and lungs. We will discuss how the ability to digest a chocolate torte changes with age. Our discussion will end with the topic of reproduction.

Along the way, we are going to discover an interesting phenomenon. We will find that certain cells in our tissues simply stop working after a while. When that occurs, some of them roll over and die. This death greatly affects the function of the tissue in which the cell resides. When enough death occurs, the tissue is impaired. And when enough tissue is rendered dysfunctional, we come face to face with the debilitation of aging.

We will further observe that these cell deaths do not happen all at the same time. Some tissues remain viable for many decades, some wear out rather quickly. In many ways it reminds me of how a house ages. Some items need attention fairly quickly – a wall gets scratched, a window becomes broken, the exterior paint becomes brittle. Some items break down only after a while – the plumbing begins to leak, wires get frayed, the roof needs to be replaced. Other items very seldom deteriorate – the fireplace or the foundation, for example. Our bodies are just like the house, with some systems deteriorating fairly quickly, while others hold up well for many, many years. Why this differential occurs, and what starts the process in the cells that are responsible for the changes are the great questions in aging research.

We will also discover something else in this section. Aging does not occur in the same way in every human being. *We will find that growing old is a very personal, very individual process.* Certain generalizations can be made, but the genetic roll of the dice dictates diversity as the norm. People's lifestyles additionally affect what will happen, an idea we will cover in the last section. I will use words like 'young adult' to describe characteristics in post-puberty youth (to around age 20) and 'later adulthood' to describe characteristics in people over age 50. All this to keep in mind that there are many exceptions to the 'rules' of aging, and when we talk of specific processes, we are also talking of generalized averages.

With these thoughts, we will embark on our tour of the Clock's nuts and bolts. Our first stop will be at a famous writer's bedroom mirror. What she observed, however, was no practical joke.

4

How the skin and hair age

'My complexion is black and white and every wrong colour,' an exasperated Jane Austen wrote in her diary. 'I am more and more convinced that bile is at the bottom of all I have suffered.'

The famous author was describing the dermal particulars of a disease that had afflicted her for two years – and would eventually take her life. Although unknown to medicine at the time, Austen was suffering from Addison's disease, the same syndrome that more than a century later, would afflict John F. Kennedy. Austen thought she was plagued with ill humors, specifically bad bile, a general pathological idea that muddied much of Western medicine in those days. It wasn't until 150 years later, in 1964, that the truth came out. A physician writing in the *British Medical Journal* made the correct diagnosis by poring over some of the remarkably detailed descriptions in her diary.

The physician found a perfect summary of Addison's symptoms, an insidious and progressive deterioration of the adrenal cortex. These organs, which secrete a variety of important hormones, sit on the roof of your kidneys in the same manner that snow sits on a mountain top. Their atrophy leads to a severe drop in blood pressure, to weight loss and an overall feeling of weakness. If left unchecked (it is usually treated with steroids), Addison's also causes an overproduction of the skin pigment melanin. This pigment causes the skin to darken considerably, producing an uneven, blotchy black/brown/white complexion all over the body.

The disease cruelly withered Austen in her late 30s, while she was at the flower of her talent and success. Born in a rector's household, she was the second daughter (and seventh child) of a family of eight. At the age of 20, she wrote her first novel, entitled *Sense and Sensibility*. By discussing the everyday happenings of very ordinary people, she gave the English novel its uniquely modern bouquet. By the age of 33, she had written the works that would forever change the tenor of English prose.

It was a persistent habit for such everyday details – this time in diary form – that allowed the English physician to discern the reason for her death.

In the winter of 1815–1816, Austen became weak and chronically fatigued. She developed a severe pain in her back. Though she kept writing, her joints began to ache and she grew feverish. She focused on treating her gastrointestinal disorders (another Addison trademark) with the purgative cures of the time. These did nothing but increase her nausea and further enfeeble a steadily weakening constitution.

Perhaps the symptom that bothered Austen most was the change in her face. By New Year's Day 1817, her coloring had changed into a socially unacceptable patch-quilt of browns and whites and blacks. The diagnosis was that this mysterious bad bile had somehow begun leaching into her bloodstream. Austen spent most of her time on a couch that winter, expending what energy she had in examining her complexion, looking for signs of disease progression. As her hands began to shake, Austen wrote less and less, and finally nothing at all. She was able to finish one last novel, *Sandition*, in the early spring of 1817. Curiously the book was a ripping satire of contemporary English health care.

The end came on July 18, 1817. Her family moved the dying author from her home in Chawton, England to Winchester, to be under the care of an expert physician. To no avail, of course. Her last words were to her favorite brother Henry, to whom she whispered: 'I want nothing but death.' At 4:30 in the morning of that day, Jane got her wish.

Skin and aging

In 1849, Dr Thomas Addison correctly explained the nature of the disease and gave the disorder its name. Also called primary adrenal insufficiency, the symptoms are known to occur because of a lack of a class of hormones called corticosteroids. Without exogenously supplied pharmaceuticals, the characteristic skin discoloration becomes an adequate measure of the progression of the disease.

In this section, we are going to use this dermal aspect of Addison's disease to discuss the effect of aging on the skin. Unlike Austen's condition, however, senescence isn't restricted to the hyperactivity of a melanin gene in certain cells. As the years go by, nearly every layer of the skin is affected by the aging process. Because these changes are dramatic, the condition of the skin is the most obvious reflection of our terrestrial tenure. The purpose of this section is to describe what happens as it gets older. We will begin with a short introduction of the anatomy of human skin and then talk about its biological maturity.

A quick anatomy lesson

As awful as Jane Austen might have felt about her complexion, human skin is really wonderful stuff. It waterproofs the body, provides a physical barrier against nasty pathogens, regulates temperature and continually talks to our brains. A typical adult has over 6 pounds of dermal material. If it were stretched out in tent-like fashion, our skin would cover an area over 20 square feet. Human skin has long been the envy of structural engineers, who marvel at its sensitivity, elasticity and durability.

Though it may not be obvious to the naked eye, skin is actually composed of many layers. There are three overall divisions, termed epidermis, dermis and hypodermis. There are subdivisions within these layers, each with their own specific functions. We will consider the structure and chemistry of some of these tissues, and describe what occurs to them as aging progresses.

The epidermis is the tissue you can see. It has cells which produce a hard protein called keratin. Although this is the material of which your fingernails are made, keratin also exists all over your body's surface area. It is the biochemical that keeps your skin from water-logging. In addition to these keratin-producing cells, there are other cells called melanocytes. Their primary function is to produce the protein melanin. Normally responsible for skin pigmentation, melanin was the molecule manufactured with such haphazard abundance on Jane Austen's face.

There are great numbers of these and other cell types embedded within the epidermis. They are continually growing, migrating to the surface and sloughing off into the rest of the world. You can see this amount simply by looking at the dust in your room on a sunny day. On average, 75% of human household dust is made up of dead skin cells. That's not surprising when one considers that by the time we are 70, we will have shed over 40 pounds of skin.

The layer underneath the epidermis is called the dermis. These groups of cells and molecules give the skin its strength (structural rigidity), extensibility (the ability to stretch) and elasticity (the ability to return to an original shape after stretching). The reason has to do with the kinds of proteins the various cells of the dermis manufacture. There is a protein called collagen, for example. It can attach to other collagen molecules like a hook to a latch. This attachment, called cross-linking, gives the skin both form and strength. Another protein within the epidermis is called elastin. As its name implies, this molecule confers the property of elasticity to the skin. After extension or contraction, the skin returns to its original shape because of the presence of this molecule. As you can see, if the cells of the dermis ever stopped producing these proteins, the skin would take on some very different physical properties. In aging, that's exactly what happens.

The third layer, underneath the dermis, is called the hypodermis (also called the subcutaneous layer or superficial fascia). It is composed primarily of an even layer of fat cells. Various fibers extend from the dermis through this layer and attach to the underlying muscles and blood vessels. The hypodermis acts as a kind of structural glue that holds the rest of the skin to the inner tissues. It also possesses nerve endings called Pacinian corpuscles. These little structures give humans their sensitivity to externally applied pressure.

There are other skin organs which extend through all the layers of the cells. There are over 3 million sweat glands, useful in regulating body temperature. On a typical summer's day they will pump out over 2 quarts of fluid; in a desert environment, they are capable of producing 2.5 gallons of fluid. There are also hair follicles, which unsurprisingly grow body hair. Attached to these follicles are tiny squirt-gun-like structures called sebaceous glands. Their primary purpose is to secrete oils to the surface, which keeps our skin from becoming rough and abrasive.

Historical attempts at modification

As you might expect, the aging process exerts its unwelcome force on all three layers of our skin. And since the skin's location is external, our dermis was historically the first battleground in our attempts to slow the inexorable ticking of The Clock of Ages.

Both sexes have used various materials to paint, muddy or alter our skin and thus change our appearance. Historically, certain make-up regimens were applied with such skill that the illusion had legal consequences. In 1770, the British Parliament passed legislation which said, in part:

> ... women of whatever age, rank or profession, whether virgins, maids or widows, who shall seduce or betray into matrimony, by scents, paints, cosmetic washes, artificial teeth, false hair, shall incure the penalty of the law as against withcraft, and that the marriage shall stand null and void.

The use of dermally applied make-up carried penalties that had nothing to do with legislation, however. Ancient Greek men, for example, would whiten their faces with a powder prior to the application of rouge. The general formula for this powder would be used for the next several thousand years to whiten European faces as well. Unfortunately, it contained large quantities of lead, which leached into their ancient and medieval bodies like acid. This caused mental impairment and complexion disfigurement and resulted in untold numbers of premature deaths.

The rouge was not much safer. Its base was gentle enough, mostly a combination of vegetable and fruit pulps. But the rouge pigment was supplemented with a red sulfide of mercury known as cinnabar. Once applied to the lips or skin, the mercury found its way into the bloodstream. How many miscarriages, deformations, stillbirths and other pathologies resulted from such heavy metal poisonings can only be approximated. These days, the substances of the make-up industry are well known, heavily regulated and very safe. But the price we paid for attempting to alter the natural biology of our skin's three layers has been historically quite heavy.

The reason for the make-up

Regardless of the make-up technology, no formula in the world can completely hide the inexorable march of the aging process across the layers of our skin. This aging occurs in different adults according to different timetables, and depends, perhaps not surprisingly, upon the skin's direct exposure to the elements. Aging changes the longevity of the cells within the skin as well as the kinds and amounts of molecules they secrete. What follows is a description of some of these processes, starting with that benchmark of dermal senescence, skin wrinkling.

Wrinkle, wrinkle little star

Wrinkling happens for many reasons. The furrows can be created by the way we move the muscles in our face. It takes about 200 000 frowns, for example, to produce one permanent brow line. The older we get, the more susceptible we are to the creation of these wrinkles. The forces of aging contribute to this susceptibility by acting on all three layers of skin (Figure 9).

Let's consider the epidermis. As you recall, dead cells are sloughed off and replaced by the living cells beneath them. As we get older, the rate at which skin cells are shed becomes greater than the rate at which they are replaced. This results in a loss of tissue. And a change in appearance. The cells that replace the absent brethren are generally less organized and the patterns that form become more irregular and uneven. This unevenness is seen with greater frequency in skin unprotected by clothing. Persons whose interests and occupations keep them outdoors for long periods of time have epidermis that is dramatically less organized in the exposed areas. This loss of cells and change in organization results in an overall epidermal fragility. It makes our skin more susceptible to creasing and folding, which, dermatologically speaking, means wrinkles.

The aging process also works within the second general layer of skin cells, the dermis. We discussed earlier the fact that cells in this layer make the protein collagen, which strengthens our skin by forming cross-linkages with each other. These linkages turn out to be very stable and are replaced more slowly, the older we get. This results in skin stiffness and the tissue loses a lot of its flexibility. The elastin fibers, which as you recall mediate extensibility, become more brittle with age. Because of this change, the ability to conform to the movements of arms and legs is reduced. Coupled with the changes in collagen, these differences mean that the skin is more likely to form furrows and lines. Since it cannot return easily to its original shape once stretched, it is also more likely to sag.

Other layers

The aging process doesn't stop at just the dermal layers. The fatty layer (hypodermis) underneath the dermis also undergoes changes with aging. The amount of fat tissue in this layer decreases with age, and its loss is

Why human skin wrinkles and sags

Human skin is subject to enormous stresses. As it ages, it is less able to handle the strain.

Our skin must conform to the pressures of moving limbs, withstand the internal motion of blood vessels and muscles and snap back to the body's original contours. Skin loses this flexibility with age. Wrinkles, sags and a loss of firmness result. Aging exerts its biological effects on all parts of human skin including the epidermis, dermis, oil glands and fat tissue.

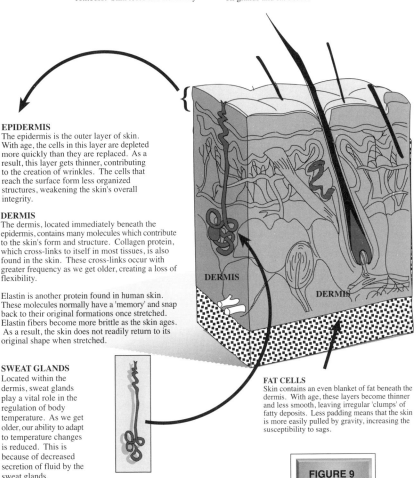

EPIDERMIS
The epidermis is the outer layer of skin. With age, the cells in this layer are depleted more quickly than they are replaced. As a result, this layer gets thinner, contributing to the creation of wrinkles. The cells that reach the surface form less organized structures, weakening the skin's overall integrity.

DERMIS
The dermis, located immediately beneath the epidermis, contains many molecules which contribute to the skin's form and structure. Collagen protein, which cross-links to itself in most tissues, is also found in the skin. These cross-links occur with greater frequency as we get older, creating a loss of flexibility.

Elastin is another protein found in human skin. These molecules normally have a 'memory' and snap back to their original formations once stretched. Elastin fibers become more brittle as the skin ages. As a result, the skin does not readily return to its original shape when stretched.

SWEAT GLANDS
Located within the dermis, sweat glands play a vital role in the regulation of body temperature. As we get older, our ability to adapt to temperature changes is reduced. This is because of decreased secretion of fluid by the sweat glands.

FAT CELLS
Skin contains an even blanket of fat beneath the dermis. With age, these layers become thinner and less smooth, leaving irregular 'clumps' of fatty deposits. Less padding means that the skin is more easily pulled by gravity, increasing the susceptibility to sags.

FIGURE 9

not uniform. This makes the underlying layer anything but smooth, the fat instead appearing to be deposited in specific areas. As a result of such loss, the fibers that run through this underpadding, connecting the dermis to the muscles, are not as stable. Coupled with a deterioration of muscle mass (and even the amount of underlying bone), this means the skin is more easily dragged down by gravity. The decrease contributes even more to the observed sagging.

The sweat and sebaceous glands are also affected. In our younger years, neural messages received by our sweat glands command the glands to secrete cooling fluid onto our epidermis. Later in life, the information flow (called autonomic input) is greatly reduced. Thus it is more difficult for humans to adapt to hotter conditions later in life. The sebaceous glands, which as you recall synthesize lubricating oil, are also affected. The overall production of oil is greatly reduced. Without their balm, the skin becomes dry, brittle and very susceptible to external abrasive forces. There is more skin damage, and since the repair mechanisms don't work as well, a dramatic change in our complexion occurs.

The body overall

Changes in the dermis and its interaction with the other layers affects the appearance of the large structures they overlay. This is most easily seen in the face (Figure 10). For example, the changes in elastin levels coupled with the shifting fat shows up under our chin. This alteration creates a dramatic sag, forming the characteristic 'jowel' or 'double chin.' The visual effect is abetted by that bone resorption mentioned earlier. Loss of bone in the lower part of the jaw makes the chin appear to shrink or recede.

This aging of the skin affects more than just our jaws, however. Our nose and ears tend to get broader and longer later in life. Fat and fluid tend to accumulate in the skin under the eyes, causing them to sag, creating characteristic 'bags.' The deposition of dark pigment in that same area gives the appearance of 'sunken eyes.'

The alterations in these three layers can be seen throughout the body. 'Age spots' appear everywhere. These are really just a gathering and enlargement of pigment-producing cells, differently arranged as the skin becomes less organized. Aging affects the number of moles on our body (they increase) and even perturbs the blood vessels that feed the skin (we

How the face ages

Examining what happens to the face says volumes about the skin's aging process.

The aging process affects all three layers of human skin, the epidermis, dermis and hypodermis. Changes in the human face dramatically illustrate many alterations experienced by these layers. Shown below left is a cartoon representation of a man at middle age. At right is a cartoon of the same man, 30 years later.

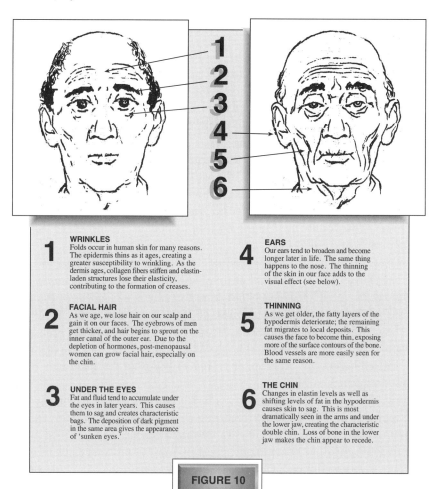

1 WRINKLES
Folds occur in human skin for many reasons. The epidermis thins as it ages, creating a greater susceptibility to wrinkling. As the dermis ages, collagen fibers stiffen and elastin-laden structures lose their elasticity, contributing to the formation of creases.

2 FACIAL HAIR
As we age, we lose hair on our scalp and gain it on our faces. The eyebrows of men get thicker, and hair begins to sprout on the inner canal of the outer ear. Due to the depletion of hormones, post-menopausal women can grow facial hair, especially on the chin.

3 UNDER THE EYES
Fat and fluid tend to accumulate under the eyes in later years. This causes them to sag and creates characteristic bags. The deposition of dark pigment in the same area gives the appearance of 'sunken eyes.'

4 EARS
Our ears tend to broaden and become longer later in life. The same thing happens to the nose. The thinning of the skin in our face adds to the visual effect (see below).

5 THINNING
As we get older, the fatty layers of the hypodermis deteriorate; the remaining fat migrates to local deposits. This causes the face to become thin, exposing more of the surface contours of the bone. Blood vessels are more easily seen for the same reason.

6 THE CHIN
Changes in elastin levels as well as shifting levels of fat in the hypodermis causes skin to sag. This is most dramatically seen in the arms and under the lower jaw, creating the characteristic double chin. Loss of bone in the lower jaw makes the chin appear to recede.

FIGURE 10

lose more skin-nourishing capillaries with age). This results in a loss of pinkish tone throughout the body. As the fatty layer is eroded, the vessels that are left become more visible. So do the bones. Varicose veins appear on our legs as small, irregular bluish lines.

The amount of time we spend on this earth is written into our skin like ink on vellum.

A hair-raising fortune

For better or worse, our terrestrial tenure is observed in other external parts of our body than just the skin layers. The passage of time can also be observed in the hairs the skin creates. In the next section, we are going to discuss a process that has affected the self-image of countless millions of men and women: the effects of aging on hair loss. So crucial is hair to many people's social well-being that millions of dollars are spent by commercial research institutes in an attempt to understand – and impede, this aspect of senescence. One of the early researchers was a handsome young American, named John Breck. Famous for his temper, his frustration with the hairy genocide on his scalp caused him more than one time to utter the following: 'Aagh!' he would shout at the top of his lungs. 'I've almost lost it all!'

The captain of a turn-of-the-century Massachusetts fire department, John Breck really *had* lost it all scalp-wise, the process nearly complete by the age of 25. The doctors told him that there was no cure for his kind of baldness, that he should just go home and accept the hand of fate. His continuing dalliances with the mirror fueled instead the opposite motivation. Refusing to listen to his physician's advice, he quit the department and embarked on a career of trying to preserve the little hair that remained attached to his scalp.

Breck's first strategy was to turn his home into a first-rate hair research laboratory. He created (useless) hair-restoring preparations. He invented and refined scalp massage treatments. He even opened his own hair loss clinic in Springfield, Massachusetts. Most relevant to his financial success, John Breck invented a series of shampoos, first for normal hair, then for dry and oily hair. These concoctions became the financial epicenter of the Breck line of hair products, a group of products still retailed today. It was a credit to Mr Breck's business acumen that this business started and flourished during the Great Depression of the 1930s.

It was a credit to The Clock of Ages that such intelligence did nothing to restore even one hair on the head of Mr Breck.

Hair today, gone tomorrow

As John Breck found out, the changes occurring on our scalp appear to be just one more example of the immutability of the human aging process. Only very recently – and only under certain conditions – have the events been affected by medical intervention. There are of course two major visual changes associated with an aging head, the change in the number of hairs and an alteration in the color of the hairs that remain. After a brief discussion of the relevant anatomy, we will examine each process.

Hair is produced in the skin's hair follicles (Figure 11). There are about 5 million follicles in an adult human being, about 120 000 sequestered in the scalp. Human hair is not alive, of course, and is therefore not capable of growing. The cells inside the follicle are alone responsible for development. Called germ centers, these cells are nourished by their very own blood vessels. When functioning properly, these germ centers create hair strands by secreting long garlands of protein at the bottom of the follicle. As more proteins are added to the end of the growing hair, the hair is mechancially pushed out through the pore.

The follicles in the scalp are quite busy. Taken together, they can produce as much as 100 feet of protein in a 24-hour period, or about 7 miles of hair a year (a typical hair in a single follicle grows half an inch a month). They don't do it all at once, however. Hair growth on the scalp is regulated in cycles. Each hair follicle will enter a growing phase for 3– 5 years, and then rest. If it loses its hair, the follicle may not replace it for three months in this resting phase. Then, for unclear reasons, it picks up speed and starts growing again. The cycles are staggered across the scalp, with about 10% of the follicles in a resting phase at any one time.

A follicle contains more than just hair-producing cells. There are cells that give an individual hair shaft its particular color. The cells responsible are the melanocytes. As you recall, these cells secrete melanin, the protein-turned-paint our body uses to give us skin pigment. And, as it turns out, hair color too. The melanocytes deposit their pigment in the root of the hair, coloring the hair shaft proteins as they are made. If pure melanin is made, you will have brown to black hair. If an analogue of melanin called phaeomelanin is made, you hair will be reddish to golden. If the

What aging does to your hair

Balding and graying are only part of the story. Here's what happens to the human scalp.

The aging process affects groups of cells within the hair follicle. One of these groups manufactures hair-shaft proteins. These cells are termed germ centers. Another group of cells manufacture proteins involved in hair color. These cells are termed melanocytes.

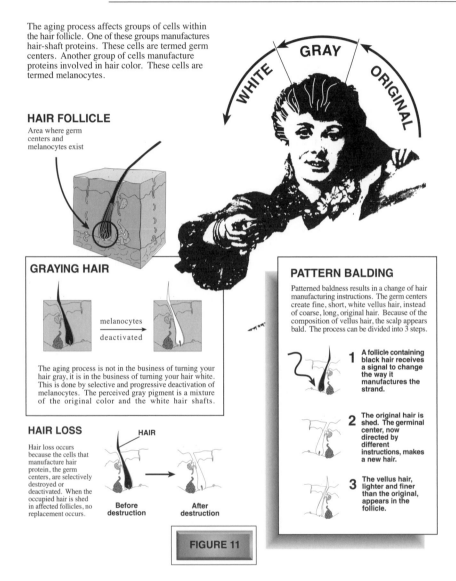

HAIR FOLLICLE

Area where germ centers and melanocytes exist

GRAYING HAIR

melanocytes
deactivated

The aging process is not in the business of turning your hair gray, it is in the business of turning your hair white. This is done by selective and progressive deactivation of melanocytes. The perceived gray pigment is a mixture of the original color and the white hair shafts.

HAIR LOSS

HAIR

Hair loss occurs because the cells that manufacture hair protein, the germ centers, are selectively destroyed or deactivated. When the occupied hair is shed in affected follicles, no replacement occurs.

Before destruction

After destruction

FIGURE 11

PATTERN BALDING

Patterned baldness results in a change of hair manufacturing instructions. The germ centers create fine, short, white vellus hair, instead of coarse, long, original hair. Because of the composition of vellus hair, the scalp appears bald. The process can be divided into 3 steps.

1 A follicle containing black hair receives a signal to change the way it manufactures the strand.

2 The original hair is shed. The germinal center, now directed by different instructions, makes a new hair.

3 The vellus hair, lighter and finer than the original, appears in the follicle.

cells quit functioning altogether, your hair will be white. As you might expect, this lack of function is what occurs in the aging process.

Hair production, whether we observe changes in color or changes in amount, can have profound effects on the course of people's professions. One of the clearest examples of the emotional bonds people have with their hair occurred in the life of the late American movie star Michael Landon. He was a track star in his early career, once setting a national record by throwing the javelin 211 feet. He even got a scholarship to a major university, on his way to a promising, perhaps professional, athletic career.

Unknown to many, Landon harbored a superstition. He attributed his athletic success to the long brown hair he groomed and washed every day. In a manner similar to the biblical Samson, he felt that strength came from length, and if he was ever to have it shorn, his career would be over. This nightmare was realized in his freshman year. Older members of the track team, in a ritual hazing event, held him down and gave him a crewcut. Psychologically devastated, Landon could only throw the javelin three-quarters of his best mark for the rest of the season. He hurt his arm, quit the track team and dropped out of school.

He became an actor instead, to the great delight of millions of people around the world.

Thinning and graying

If Michael Landon had known there was a daily, natural loss of hair, he might not have been so paranoid. The average human adult, for example, loses between 70 and 100 hairs per day. This is true whether you are a male or a female. As you may be aware, the amount of loss per unit time can be changed by diet, illness, stress and other factors.

During our younger years, the replacement rate equals or exceeds the rate of loss. As we experience the aging process, this delicate balance changes, and we don't make as many hair shafts as we lose. Why does this occur? In later years, The Clock of Ages commands the germ centers in some follicles to shut down. When these centers are rendered dysfunctional, no hair proteins are made and we experience hair loss. While this loss is usually associated with males, the destruction of the germ centers occurs in both men and women.

Except in one case. There is a phenomenon called male patterned baldness, which has a genetic basis and occurs, consistent with the name, in men. This process is different from age-related hair loss in that the follicles are not destroyed. Instead, they are mysteriously given new instructions on how to manufacture hair. The follicle is told to stop making the coarser normal stuff and instead make something more delicate. The hair coming from the new instructions is a very fine, nearly invisible kind of protein called vellus hair. This hair is not easily seen by the human eye, and thus the person appears to be balding. He's not really, his scalp has just changed the kind of hair he creates.

This change nicely illustrates the fact that aging doesn't always wage a war of annihilation against our hair. For reasons not clearly understood, we can just as readily sprout hair – in sometimes embarrassingly unexpected places – as we can lose it. In men, the hair on the eyebrows may get longer and coarser with age. Males can even grow hair on the inner surface of the outer ears. Women may develop hair on their faces, specifically on the chin. This new growth may be related to the loss of specific hormones after menopause.

Graying

A universal sign of aging is not just hair loss, but a change in hair color. We commonly – and mistakenly, it turns out – refer to this recoloration as graying. The extent of recoloration as well as the age of onset varies from individual to individual and is described next.

Though it may not appear obvious, there is really no such thing as gray hair. The aging process dictates that after a period of time, our hair is to turn white, not gray. The grayish hues we perceive are just the intermediate steps as this alabastering unevenly advances across our scalps. The amount depends on how much of our original color mixes with the white. While universally experienced by both sexes later in life, the recoloration is not totally an age-related event. Many people turn gray and then totally white long before they achieve middle age.

The biochemistry of graying

To understand the biology of hair recoloration, we once again must turn to the cells that mediate skin and hair color, the melanocytes and to their molecular tempera, melanin. As we get older, there is a decrease in the number of functioning melanocytes. Less and less pigment is secreted into the base of the hair follicle as a result. The hair loses its original color and becomes increasingly lighter in hue. Eventually, the number of melanocytes still secreting pigment decreases to zero. When that happens, a pure white shaft of hair is created. As time goes by, the follicles still creating hair have all turned into non-pigmenting producers. Thus the proper term is not hair recoloration, but, more accurately, hair decoloration.

The alterations senescence makes to the skin are some of the most externally obvious signs of aging. They are experienced in our morning rituals, when we look in the mirror and visually inspect our bodies for physical changes. We observe the wrinkles, look at a receding hair line, feel the bulging gut and examine the cellulite. As senescence deploys its powerful forces throughout all our skin layers, we become engaged in a battle with, of all things, a calendar. This conflict has caused depression. It has inspired scientific creativity. It has initiated multi-billion-dollar businesses in both the cosmetic and the pharmacological industries. It has even caused a famous author to accurately catalog the symptoms of a disease that had not yet been described.

And, so far, we have found absolutely nothing that can stop it.

The motion for a change

Another obvious sign of aging is experienced in the morning, although it has less to do with looking in the mirror than with getting out of bed. This has to do with the interactions of our muscles with our bones and joints. Though we mostly observe the changes we see in our skin, the alterations in our ability to move are also experienced. Anyone who has undergone the aging process knows how easily this can occur.

In the next chapter, we are going to discuss how the coordination of joints, muscles, bones and ligaments interacts with The Clock of Ages. We will examine how alterations in underlying tissues changes our ability to move through physical space. We're not going to start with a medical anatomy lesson this time, but with a medical history lesson. Specifically, we will discuss a woman who knew a lot about human bodies, and in her last years, a lot about human motion.

5

The aging of bones, muscles and joints

'You what?' her mother screamed across the dining-room table. The fire in her mother's voice matched the fire in the nearby family hearth. 'How could you possibly bring ruin on the name of this good family?' A young Florence Nightingale outwardly cringed at the outburst.

'I will not have my daughter taking on the role of a chamber-pot maid!' her father roared in response to his wife. 'And in a hospital, for God's sake!' He stormed out of the room, his dinner untouched. Florence, suffering the steely glares of the rest of her family, grew short of breath. She began to feel weak.

'See what you've done to your father . . .' her mother started. And then, noticing her daughter's sudden frailty, 'And here we go again, with yet another famous dizzy spell. If you cannot stand up to your family, how will you stand up to a doctor?'

Florence Nightingale stumbled out of her chair and virtually crawled to her bedroom. Since she had announced to her wealthy family her intentions of becoming a nurse, there had been immediate and non-stop conflict. This shortness of breath and heart palpitation had started just as suddenly. Socially, nursing was the closest thing to low-born slavery the elder Nightingales could conceive. To Florence, it was the only thing she wanted to do.

Fortunately for the rest of the world, her family's objections only solidified her resolve. In 1853, at the age of 33, she left home to practice her profession. Her weakness and palpitations ceased almost immediately. She learned the tools of her trade and grew in confidence. By the time she left for the Crimean War in 1854, Florence Nightingale was a force of nature.

Her accomplishments as the founder of modern nursing are legendary, of course. Her selfless devotion to the injuries of British soldiers, her almost single-handed transformation of nursing into a dignified and honored profession, have made her an international symbol of courage and compassion. Curiously, this bravery and determination generally refer only to the 632 days she spent in the service of The Crown. When she returned to England, the symptoms of anxiety she had experienced in her home came back as well. Florence Nightingale immediately became bedridden. And she would remain that way for the next 54 years.

The power of immobility

The reasons for her illness remain mysterious. At the time, she was diagnosed with dilatation of the heart and neurasthenia, which was then defined as

> A condition of weakness or exhaustion of the nervous system, giving
> rise to mental and bodily inefficiency.

It was essentially a non-diagnosis. Modern clinicians have turned increasingly to psychology, rather than to somatic pathology, to explain this five-decade prostration. The illness, though perhaps real to her, may have been the result of a productive hypochondria. It did not affect her mind at all. She continued to create earthquakes within the English medical system, founding a school of nursing, supervising public health in India, publishing a treatise on hospital sinks. Nightingale regularly had audiences with generals, cabinet ministers, doctors and viceroys, all seeking her advice, and all sitting round her sickbed to hear it. Such attention made her extremely difficult to live with. Nightingale became bossy and short-tempered, and learned to carry decades-long grudges towards those who crossed her. She died at the age of 90, still issuing orders and still under the covers. Dr George Pickering, a 20th century physician who has championed the psychoneurotic theory of Nightingale's life, said, 'Of one thing I am quite sure, I should not have liked to be the doctor who tried to explain to Miss Nightingale the nature of her illness.'

Deep in our bones

It is one thing to imagine an immobility with roots in an attitude, quite another to experience one with roots in the aging process. As we get older, changes occur in all the structures we normally use to subvert the forces of earth's gravity. And most of these changes are not rooted in psychopathology but in more definable physiology. In this chapter, we are going to describe the effects of The Clock of Ages on our musculoskeletal system. We will start with events occurring in our bones, and then move to muscles and joints.

If our skin is a model of external flexibility, our skeletal structure is a model of internal rigidity. Its major function is to keep us from looking like giant amoebas, allowing us dextrous mobility while protecting vital organs. To accomplish this task, bone is made of three substances:

(1) Mineral deposits, principally calcium (taking up about 45% of its volume);
(2) Soft tissue (cells and blood vessels, about 30% of the volume); and
(3) Water (25%).

Engineers have often marvelled at bone's twin abilities to be extremely strong, yet provide lots of pliability. Human bones can withstand pressures of about 24 000 pounds per square inch, about four times that of reinforced concrete. Yet if you removed the mineral deposits, what you had left would be flexible enough to tie into knots.

There are several kinds of bone scattered throughout our body, a fact which complicates a description of how human skeletons age. There are compact long bones (in our thighs and arms), spongy short bones (wrists and ankles), flat bones (ribs and skull, with spongy material in between them) and weird irregularly shaped bones, which can either be compact or spongy. The kinds of events that occur during aging differs depending on whether the bone is made of compact material or spongier, lighter material.

The skeleton, though extremely strong, is hardly a static don't-ever-bother-me-again structure. The adult body is continually resculpting its internal scaffolding, incessantly chiseling away at older structures and replacing them with newer bone. The physiological artists are actually cells that come in one of two overall classes. The first class of cells, termed osteoclasts, generally tear down (resorb or demineralize) solid bone material. The other class of cells, termed osteoblasts, do the opposite. They gather and apply new bone material to our skeleton. These cellular sculptors are so busy that they completely replace the entire human skeleton about once every seven years. Their inability to keep up this work in later years accounts for much of the skeleton's aging process (Figure 12).

The structure of human bone

The internal structure of bone has many components.
Here's what a few of them look like.

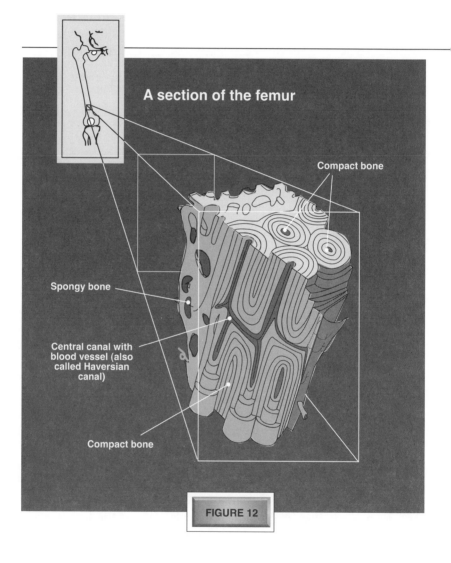

A section of the femur

Compact bone

Spongy bone

Central canal with
blood vessel (also
called Haversian
canal)

Compact bone

FIGURE 12

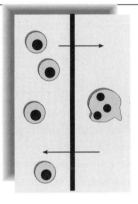

OSTEOBLASTS
make bone

OSTEOCLASTS
erode bone

THESE ACTIVITIES
remodel the skeleton

Osteoblasts are round cells whose primary function is to create bone (a process termed ossification). The oblong shapes inside the cells are mitochondria, intracellular organelles which supply this bone-building activity with energy. As described in the text, human embryos first create a skeleton made of cartilage. This will eventually be replaced with solid bone. In the adult, osteoblasts work with osteoclasts in bone remodeling, a term describing the replacement of old bone tissue with new.

Osteoclasts are giant, multi-nucleated cells often found on bone surfaces. These cells are responsible for a process known as bone resorption (resorption means the loss of a substance because of physiological, or even pathological processes). Osteoclasts extend cellular 'arms' into the bone and deposit acids and enzymes. The subsequent caustic reaction results in bone erosion, with excess calcium dumped into the blood stream. Osteoclasts are particularly numerous in bone depressions and may contribute to their formation.

In the human adult bone, a delicate balance is maintained between the action of the osteoblasts and osteoclasts. If the osteoblasts are too active, the bones will become abnormally thick. Surplus calcium can form uncomfortable spurs and bumps, resulting in decreased mobility. If the osteoclasts are too active, the bone will become thin and brittle. This loss of tissue creates weaknesses which can result in bone breakage. This delicate balance is altered with age. Demineralization of the bone and vulnerability to fracture result.

The amount of inorganic material upon which these cells work is best seen in the products of cremation. The modern furnaces employed are so hot that most of the organic components – and the water – are destroyed or evaporate in the first hour. What remains is calcium, lead, zinc and potassium in the form of white ash. The majority of these compounds come from the bones of the deceased. These are then crushed and usually placed in an urn. Thus, what was never 'alive' inside your body when life existed is now the only thing left when life is gone. Lately, there have been some enterprising Americans who, for a fee, will put those ever-permanent ashes in a stuffed animal – and you can have several choices. Most popular is the teddy bear.

Low-interest bones

The aging process in bones goes a lot more slowly than the events in a crematorium, and it performs exactly the opposite function. Essentially, there is a progressive loss of bone mineral content as the years go by. The rate of the 'chiseling' by the osteoclasts begins to exceed the rate of replacement, which simply means there is an overall reduction in bone mass. The result? The bones become structurally weaker. Their main task is to resist mechanical pressure, and, as we get older, they begin to fail.

The tug of war between these two cells is not the entire story, however. There appear to be age-related delivery problems in the ability of the small intestine to continue to absorb some basic building blocks (calcium, for example). This failure may be due to a depletion in vitamin D. This vitamin is involved in snatching calcium out of the food you eat and giving it to systems that can feed your bones. Without an adequate supply of these raw materials, new bone minerals would not be laid down. And the bone would get weaker and weaker. There may also be an increase in the overall porosity of a given bone's internal structure. This would have the same weakening effect.

Regardless of the cause, the failure of the skeleton to fulfil its job description is dramatic and can be demonstrated in many ways. Researchers have used as an example the ability to repair a fracture (Figure 13). When we are young, our bones break in the same way a young branch does when it is snapped from a tree. A great deal of bending must occur before it breaks, and when it does, many energy-absorbing fracture lines are created. This lack of a clean snap is fortunate. It is much easier for

A tale of two FRACTURES

When bones break in our youth

When the bones of younger humans fracture, the break is generally not very clean. There are many cracks and the primary break is often incomplete. This is due to the high proportion of fibrous material within the bone and the bone's overall reduced porosity.

The increase in the number of fracture lines provides many surfaces which can easily be rewoven. As a result, breaks in younger people tend to heal more quickly than in older people.

When bones break in our maturity

When the bones of older humans fracture, the break is often smooth and even. This is due to the absence of fibrous material in the area of fracture. It is also due to the increased porosity of the bone material, illustrated on the right. This loss of material effectively reduces the level of energy the bone can absorb before breaking. The ease of fracture is also due to the increase in 'cement lines,' which are etched remnants of sites of previous remodeling. These characteristics can greatly impede the healing process in older people.

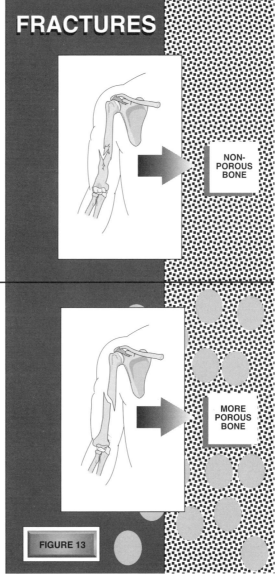

NON-POROUS BONE

MORE POROUS BONE

FIGURE 13

the body to repair a partial break with lots of little fracture lines than a hard clean break, one that might require external resetting.

When we are older, the change in mineralization makes our bones not only weaker, but also more brittle. When bones in this condition break, it is like snapping an old, dry twig. The break is generally clean, with very little flexibility to resist the fracturing force. Such accidents are unfortunate. Not only is the kind of break harder to repair, but, as we'll see later, our ability to fix them lessens with age.

The role of gender

It has been known for some time that age-related gender differences exist in the aging process. These differences show up dramatically when one examines overall bone strength. There are two ways to measure the difference. First, in a measure of total skeletal mass, men lose 3% of their skeletal weight per decade. Women lose 8% per decade. As a measure of total mineral density throughout adult life, the total reduction in women is about 30%, and in men, about 15%.

These figures represent averages that occur from beginning to end. The loss is hardly a smooth, regular progression, however. For both sexes, the rate of bone loss begins slowly (typically starting at age 39), and then begins accelerating. Women eventually achieve a rate of loss that is double that of men. Post-menopausal women are affected the most, probably due to sudden estrogen loss. There is evidence that the degeneration slows down by age 70.

The demineralizing process in aging bones is so predictable that some researchers are questioning the definition of osteoporosis as a separable disease. It may simply be the most pronounced rendition of a natural process. This idea is supported by evidence that the bone density in patients diagnosed with osteoporosis are not different from people who don't have it. Moreover, severe bone loss is not always associated with the presence of clinical symptoms. Directions in osteoporosis research have been widened by the isolation of a gene that may contribute to the symptoms. I'll have more to say about that gene later.

Taken together, the complex interplay between nutrient uptake, bone cells and mineral levels all contribute to the degeneration of skeletal tissue. As yet, modern research has not uncovered a fixed, unassailable erosion time-table that can predict its onset and severity. In a later section we will address the effect of personal habits – such as exercise and diet – on slowing down such processes. But for now, we must settle for the fact that The Clock of Ages ticks irreversibly inside our bones. This is as true for the most active heroes of this century or the most famous ones of the last.

Even if they choose to spend their final 50 years in bed.

Joint custody

In the previous section, we described the factors that contribute to the overall strength of our skeleton, and our ability to remain mobile in the face of the aging process. The primary factor contributing to our mobility has to do with the intersection of the bones in our bodies, our amazingly flexible joints. The primary factor contributing to our loss of mobility as we get older has to do with how the aging process affects our joints. The reasons for such loss is the subject of this section.

To discuss how aging affects our joints, it is important to comment on joint anatomy. Joints can be subdivided into many categories, based on certain common characteristics. One way is to organize them according to their ability to move and in so doing we find three overall classes (Figure 14): diarthrotic joints – joints that are freely movable (knees and shoulders are examples of this kind of joint); amphiarthrotic joints – joints that can move slightly (the discs lying between bones in the back are one kind of amphiarthrotic joint); and synarthrotic joints – joints that do not move (the human skull is made up of individual plates held together by connective tissue. Though they are joints, they cannot move).

Many kinds of tissue are responsible for the smooth functioning of these human joints. They are divided into structures we term ligaments, tendons and cartilage. And not just tissues. There are cells that secrete fluid into specific areas between bones. This liquid is called synovial fluid. Not surprisingly, those joints possessing liquid-accommodating spaces between the bones are called synovial joints. The joints in your knees are examples of synovial joints. We will examine the structure and fate of ligaments, tendons, cartilage and synovial joints individually.

Ligaments are cylindrical fibers made of dense, regularly patterned collections of connective tissue. This tissue is composed of relatively few cells and lots of extracellular landfill, created by those very cells. This landfill (also called 'matrix') primarily consists of the proteins collagen

Three types of human joints

All human joints undergo the aging process. What a particular joint will experience depends on its type.

There are many ways to organize human joints. One way to classify them is according to mobility. Illustrated below are the three major kinds of joints in the human body, based on their ability to move.

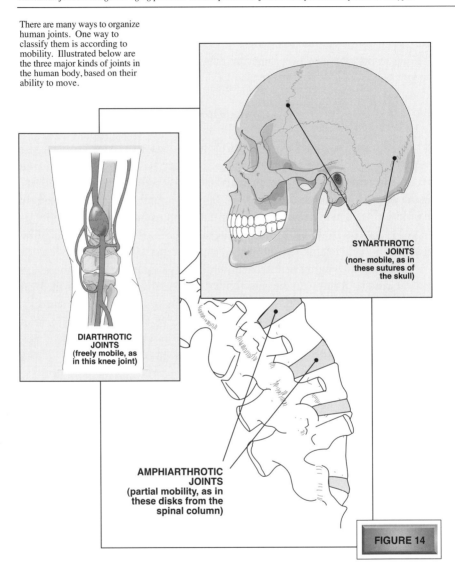

DIARTHROTIC JOINTS
(freely mobile, as in this knee joint)

SYNARTHROTIC JOINTS
(non- mobile, as in these sutures of the skull)

AMPHIARTHROTIC JOINTS
(partial mobility, as in these disks from the spinal column)

FIGURE 14

and elastin, which are responsible for the toughness of the structure. And they have to be. Ligaments connect bone to bone and, along with the tendons, regulate controlled movement.

Tendons are cords of connective tissue that attach a bone to a muscle. These tissues are also composed of the familiar collagen and elastin. Some tendons (like those in the wrist and ankle) are surrounded by tubes of very tough fibrous connective tissue. These are called tendon sheaths. Between the tendon and its sheath lies a pool of fluid, which, as you might guess, is also called synovial fluid. The sheath and accompanying structures allow tendons to slide easily. They also keep them from slipping out of place.

Most of the bones in our movable joints contain a substance known as arterial cartilage. This substance forms a transparent but extremely tough protective coating over the ends of our bones. This helps reduce the over-all friction within the joint. It also keeps the bone edge from being eroded away by the constant apposition to its partner bone on the other side of the joint. There are cells that continually secrete arterial cartilage, creating new layers as the old is worn away.

Aging the joints

As important as the joints are to our ability to function as we get older, there is surprisingly little empirical data on their aging. It is thus difficult to assess the biological restrictions of mobility, flexibility and movement as the years go by. The information that *is* available points to a tendency gymnasts have understood for decades: there is peak flexibility before your 20th birthday, and, after that, there is a steady decline in joint function (Figure 15).

As we age, tiny but very important internal molecular interactions within collagen and elastin fibers begin to change. This causes the once beautifully contiguous fibrous structures in our ligaments and tendons to fragment. This fragmentation means that the ligaments and tendons are less able to handle the stressful forces placed upon them by movement. They become less resilient and more prone to injury. As damage increases, tough, inelastic scar tissue forms in these structures, as well as unwelcome deposits of wandering calcium atoms. These changes result in further injury, crippling the flexibility and elasticity of the tissue. This causes increasing difficulty in movement.

How human joints age

As the years go by, our joints become less flexible. Here's what happens.

Illustrated below is an idealized synovial joint (named for the characteristic presence of fluid). As aging occurs, various tissues within the joint begin to break down. Loss of flexibility is due to the combined deterioration of tendons, ligaments and cartilage.

Degenerative changes are usually associated with advancing years. The processes mediating those changes actually occur before skeletal maturity is reached, however. The effects described here can occur at any time during adulthood.

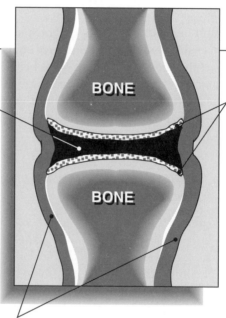

SYNOVIAL FLUID
This fluid, not available in all joints, is used to cushion the force of movement between bones. In human joints, the fluid gets 'thinner' with age. This change has a dramatic negative effect on joint flexibility.

BONE

BONE

ARTICULAR CARTILAGE
These clear structures normally provide a protective coating against friction (from bones rubbing against each other). This cartilage becomes opaque, cracks and frays as the joint ages. These changes result in increased joint pain and decreased mobility.

TENDONS AND LIGAMENTS
Changes in collagen and elastin production result in fragmented tissues. This degeneration is augmented by scar tissue deposition and calcification, primarily due to age-related injury. The resulting structures are less resilient, with increasing loss of flexion and extension.

INJURY
As joints are subjected to extreme daily trauma, increasing age results in cumulative injuries. Repairs are generally detrimental to optimal function. Fibrous and calcified new material replace elastic and resilient original tissue. This results in achiness, stiffness and a reduction in effective movement.

FIGURE 15

But more than just tendon and ligaments begin to change with age. As we get older, the arterial cartilage, that transparent friction-reducing coating on the edges of our bones, begins to change. Starting in our 20s, its color shifts from clear to an opaque yellow. Some researchers believe this alteration signals a change in its protective hardness. As we get older, the cells that normally manufacture these cartilagenous layers begin to change their molecular output. The result is that the structure becomes thinner, less springy and somewhat pockmarked. Even with normal environmental stress, extensive cracking, shredding and unraveling are observed in aging tissue. The bones underneath this cartilage, which are just as subject to gravity as in younger days, begin to wear away. The end result is that mobility is restricted.

Many researchers believe that the same age-related changes in collagen and elastin output observed in skin cells also occur in the joint. Human joints, which heavily depend on the correct amount and formation of these proteins, suffer the effects of a system-wide change. As we'll see later, dramatic alterations occur in our circulatory system's ability to bring blood to every tissue that needs it. This failure of the blood supply may have an important effect on the cells that normally make the collagen and elastin.

The ideas linking cellular function with joint deterioration, while intuitively obvious, are not greatly supported in scientific literature. Indeed, the most commonly held idea about joint aging has been the notion of 'deterioration through use' rather than 'deterioration through loss of gene function.' Blaming some amorphous effect of time for definable biochemical processes is an idea this book seeks to destroy, however. While it's true that joints are the focal point of gravity's conflict with upright movement, it's also true we hold up well against the onslaught for many years. It is the change we seek to understand.

Muscles

Whenever I think of joints, I immediately think of the muscles that have to support and utilize them. And then I start thinking of the amazing capabilities our body possesses in confronting the gravitational stresses of our planet. For some people the normal everyday stresses are not enough. There are people who have to train their muscles to tackle events we might consider extremely hazardous. One such event occurred in Romsey,

England in 1984. The act was an amazing example of the power of the human musculoskeletal system.

It was, by all accounts, a very fairly dank day in February. An unusual kind of scaffolding could be seen in a field on the outskirts of town. It caused quite a stir, because the structure was supported by a single man underneath it, balancing the scaffolding like Atlas upholding the world. Beside him was a large pile of tires and, beside them, a group of people. Their human task was singular; it was to collect those tires, climb a ladder, place the tires on the scaffolding . . . and then get out of the way.

The man upholding this structure was named Gary Windebank. He was trying to set a world's record for supporting weight – in this case tires – and was cumulatively succeeding. His associates placed 50 tires on the scaffolding above him. He did just fine. They placed 60 tires on the scaffolding. Again, he held up fine. Seventy tires. Eighty. Ninety. Ninety-five. It wasn't until they got to 97 that Gary gave up.

His total weight? 1.440 pounds of tires. Ninety-six Michelin XZX 155X13, to be specific. He accomplished a record that has been unbroken for more than ten years.

This kind of feat occurred because of the remarkable power and construction of human muscles (and, some might say, the remarkable vacuity of the intellect which can drive them). And not only power, but their almost miraculous resiliency. Human muscles are made of tissues that can retain much of their strength even into our seventh decade of life. In this next section, we are going to discuss some of the biology of human muscle and attempt to understand what happens to these tissues in our older years. We will begin with a description of our muscle fibers and then discuss how they age.

Muscle sprouts

There are three types of muscle tissues in the body. They are:

(1) Smooth muscle tissue. These muscle tissues are concerned with vascular, gastrointestinal and reproductive processes. They are usually located in the lining of organs. This variety of function has something to do with their location. The inner walls of the stomach are made of smooth muscle. So are the walls of the uterus, blood vessels and intestines. There's not much you can consciously do to control

the action of smooth muscles. In the parlance of neurologists, they are said to be under involuntary control.

(2) Cardiac muscle tissue. The cells that form this tissue make up most of the weight of the heart. By interacting with a specialized set of nerves, cardiac muscle controls the heart beat. Like smooth muscle tissue, cardiac muscle is not under voluntary control.

(3) Skeletal muscle tissue. These are the tissues that attach to bones and, with the aid of ligaments and tendons, allow us mobility. They are under voluntary control; that is, we consciously tell these muscles where to go and what to do. Our discussion of muscular aging will focus primarily on skeletal muscle tissue.

What skeletal muscles do

The job of skeletal muscle is to abbreviate their length for brief periods of time, a process called contraction. Because they are attached to bone, when they shorten themselves, they can take the bone with them. The action produces movement. When they relax, the muscles become longer once again. To achieve movement, a group of muscles on one side of the bone contract and, at the same time, a group on the other side stretches out. It's a kind of lever action, with joints serving as biological fulcrums.

There are several kinds of contraction available even within a single muscle. There are groups of tissues that are called 'fast-twitch' muscle fibers; these are useful for creating rapidly accelerating – and quite powerful – muscular contractions. These are the events normally associated with 'strength,' such as demonstrated by the tire lifter. But there are fibers capable of other kinds of contraction. One group is called the 'slow-twitch' muscle fibers. This group is involved in creating and maintaining contractions over a long period of time. The fibers that mediate our posture are slow-twitch fibers, for example.

Whether muscles are relaxing or contracting, twitching quickly or slowly, there must be a great deal of coordination between the various muscle groups in order for human motion to occur. That means heavy-duty communication, of course; physiologically speaking, that also means nerves. Most human skeletal muscles are well supplied with nerves that link directly to the brain. There are nerves that spring from a muscle and then turn right around and come back to it, however. These loops mediate our reflexes, and are not under conscious control.

The cellular connection

The command structure of nerve-talking-to-muscle and vice versa is very important, not only for the motion of the body, but also for the life of the cell. If the muscle does not receive commands to move, the muscle will simply sit there. If the communication is cut off, however, the muscle will begin to atrophy. The fibers will be cannibalized for other uses inside the body and the function mediated by the muscle will be lost. These tissues are prime examples of the 'use-it-or-lose-it' philosophy of many biological processes. As we'll see below, this unpleasant inventory control is at the center of the aging of the skeletal muscle system.

Before we describe how muscles change in later adulthood, it's important to review several aspects of their internal biology, namely their two-way system of food supply, their waste disposal, their ability to use energy and their internal construction. We will begin by reviewing their two-way system of food supply and waste disposal.

When the muscles are active, they will need nourishment – and, if the activity is vigorous – lots of it. The mobile cafeteria muscles use is the ubiquituos blood supply, carried by that intricate system of tubular highways that is our vasculature. From it the muscles receive both nutrients and oxygen. When finished munching, they secrete harmful waste products. Since the red stuff freely communicates with the body's filtering system, it is in the blood's job description to take away the muscular sewage too. This is a two-way system. There are capillaries that communicate with every fiber to make sure that it is both 'supplied' and 'cleaned.'

What a skeletal muscle cell looks like

All this feeding and waste treatment activity occurs at the level of the individual cell. If you tease away a large muscle and then view it under the microscope, you will find millions of cylindrical cells that look somewhat like they've been painted by the great Spanish painter El Greco. They are pale and elongated and lie parallel to each other. Called myofibers, these cells have multiple nuclei within their cytoplasm (most cells only have one) and come in a variety of lengths. Some are so tiny that hundreds of them could fit on the head of a pin. Some are almost a foot long, which in cellular terms is ridiculously enormous (Figure 16).

Human skeletal muscles

Here's what they look like and how they move:

When a skeletal muscle is dissected, thousands of elongated, log-like structures are observed. These structures are really giant, cyclindrical cells called myofibers. They are laid parallel to each other and are wrapped in a membrane called the sarcolemma. Many nuclei exist within these large cells, all sharing a common cytoplasm. Inside these cells are also many mitochondria, needed to fuel the muscular contractions that make human motion a reality.

Contractions occur because of the presence of tiny proteins within the myofibril. They are arranged in highly ordered stacks. Acting like molecular ratchets, these proteins alternately hold on to and let go of their neighbors. This on/off binding creates a pushing and pulling capability. Such activity is the basis for human muscular contractions.

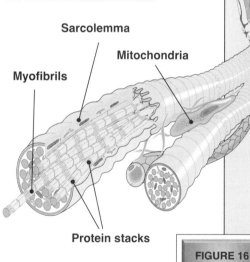

Sarcolemma

Mitochondria

Myofibrils

Protein stacks

The arm at the top has a rectangle through its upper third. What the muscles of that arm look like in cross-section are drawn and extruded here.

FIGURE 16

They are also ridiculously strong. There is on record a human being who was able to life a wheelbarrow upon which 8275 pounds of bricks had been placed. He was able to move it 243 feet. But even this record pales in comparison to the muscles in other organisms, many of which have the same overall design. There is a beetle in the *Scarabaeidae* genus so strong it can carry a burden 850 times its own body weight. If the person who moved the bricks had the same talent, he would have moved a wheelbarrow carrying almost 68 tons of bricks.

It's what's inside that counts

Whether they are in insects or humans, muscles perform work because of proteins that sit inside skeletal muscle cells. Under the microscope, the human ones look something like tiny oarsmen in an ancient Roman ship. When muscles contract, these proteins 'row' past each other, exerting a force and allowing contraction to occur. Though it is hardly the point of this chapter to describe in detail these proteins, their presence dictates whether a particular muscle is going to move a bone or not. And in aging, that is all that counts.

As mentioned earlier, the muscles receive nutrients and oxygen from the blood supply. But that has to be turned into useful energy in order to mean something to a myofiber. This is where the cell's mitochondria come in. As you recall, mitochondria are the powerhouses for most living things. This is just as true for myofibers. These little dynamos take the molecules given by the blood and turn them into the raw energy muscles need. Without this conversion, we would never be able to move.

Taken together, there are many requirements to keep a muscle functioning normally. It has to be attached correctly, it needs to be in constant communication with its command centers, it needs food supplies and waste disposal. In order to perform work, the molecules must slide past each other – and they can do this because they carry their own subcellular batteries. If any of these supporting processes degenerate, the muscle will be less and less able to carry out its normal functions. As we'll see next, The Clock of Ages affects not only tendons, ligaments and joints, but also the very stuff of the muscle itself.

The aging process

To understand how muscles are affected by the passage of years, it is important to understand how their various talents are measured. That's not easy. The reason is that muscles are so amazingly sensitive to environmental input. And they can do so many things. Does the person being measured get a lot of exercise? What is their gender? What kind of diet do they have? Does one wish to measure static (isometric) strength or active (dynamic) strength of a particular muscle group? The number of variables that must be addressed before aging generalizations can be made is enormous. Not surprisingly, there is a fair amount of controversy about what happens to our muscles in later adulthood.

A few general observations can be made, however. There doesn't seem to be a measurable loss of strength until we get to our fourth decade of life. Even between our 40s and 50s the loss is fairly minimal. By the time we reach our 60s and 70s, there is a 10–20% reduction in total strength, which, considering concurrent alterations in other body systems, isn't much at all. Where we really experience loss is in our 70s and 80s, typically 30–40% of the total. This loss is more severe in the muscles of the legs than in the hands and arms.

An example of the toll

The effects of aging on our musculature is most easily seen by examining the ages of people in athletic competitions. The oldest person ever to win a gold medal at an Olympic event was 42 years old. And he did it in 1920. The oldest person ever to play NBA basketball was 41. The oldest person ever to hold a boxing title was 38 years old. The greatest age at which anyone broke a world track and field record was 41, and this was done in 1909. Muscles are some of the most age-resistant tissues we possess. But there is a marked decline in our muscular abilities, one that cannot be completely halted by an exercise regime (see Part Three).

Not only is there an overall loss of muscular strength, there is also loss in the kinds of muscular activity we are capable of performing. For example, we tend to lose our 'fast-twitch' abilities much faster than our 'slow-twitch' abilities. As you recall, fast-twitch mediates rapidly accelerating, powerful muscular contractions. Slow-twitch creates and maintains contractions over a long period of time. This change has an interesting net affect: it allows our body to remain upright in a gravitational field and at the same time, decreases our ability to manipulate the objects within it.

What happens

Given that there are a few measurable changes that appear in most everyone, why does such a loss of function occur? And why is the loss so pliable that lifestyles dramatically affect our ability to measure it? The capacity to influence the ticking of The Clock of Ages will be discussed in a later chapter, but researchers delving into exactly what occurs in our aging muscles have found some reliable data.

Our muscular abilities decline with age, primarily because they lose some of their mass. It appears that, after a period of time, certain muscle fibers start to atrophy. They don't all do it at the same rate (hence the differences seen in our fast- and slow-twitch capabilities), and different muscle groups show different levels of destruction. But myofibers eventually die, and the more fibers that are destroyed, the less mass exists within the muscle. The cells are eventually replaced with connective tissue and, later, with fat. This reduction in mass means a loss of cells carrying those 'Roman oarsmen' molecules we described earlier, and an overall decrease in strength.

Because this loss is measured at the cellular level, the next question is: why do the cells die? The real answer will ultimately depend on what cell in which muscle is being discussed, but several hypotheses have been put forward to explain the loss of various types of muscles. Summarized below are two ideas.

The first idea takes its cue from what happens to muscle when it is unused. When these tissues no longer receive neural input, there is pronounced atrophy and reabsorption of left-over muscle molecules. There is evidence that some myofibers are reabsorbed, not because anything is wrong with *them*, but because something is wrong with the nerves that are supposed to tell the muscles what to do. When these nerves die, the muscles no longer contract. The body's molecular vultures, sensing the inactivity, begin selective destruction of muscle tissue.

A variation on this idea has to do with a loss of blood flow to a particular area. As we age, it is increasingly difficult for the blood supply to nourish and cleanse some of our deeper tissues. If this happens to a muscle, the muscle cells will die. This has the obvious effect of weakening the overall muscle tissue in which such degeneration occurs.

The second idea has less to do with communication than with energy sources. As we discussed earlier, mitochondria are responsible for giving muscles their energy. There is some evidence that they have less of an ability to pump energy into the tissues as we increase in age. This lack of energy would render some of the fibers inactive. And that indolence would lead to atrophy and cell death.

Credibility

Though it is clear that muscles can atrophy if they are not active, exactly how this works in the aging process is a matter of controversy. Even the two hypotheses mentioned above are not universally accepted in all research quarters. The variables we discussed earlier must be taken into account even at the molecular level. And even if these hypotheses provide a partial explanation, they really only push the question of aging to another level. For example, what causes the nerves to die as we get older? They maintained their health for many years – indeed, even in our sixties we can still retain remarkable muscle strength. But a change occurs. What causes the mitochondria to die? Their ability to apply a useful energy system was working when we were still in the womb. Many questions remain to be answered before we understand the effects of the years on our musculature.

What do we make of appearance?

Whether we are talking of bones or joints or ligaments or muscles, it is clear the effects of aging dramatically affect our static and dynamic appearances. There seem to be mysterious signals that simply show up at certain times and tell cells to quit doing their normal adult functions. When the delicate balance between bone creation and destruction is tipped in favor of mineral loss, our bones become more brittle. When the cells that make a certain kind of cartilage in our joints stop working, our flexibility is diminished. And when muscles stop receiving enough neural input or energy to carry out their primary tasks, they lose their mass. Ultimately, they can turn into connective tissue and fat.

What are we to make of this seemingly deliberate plan for our destruction? How do we account for the common and specific timing we all seem to experience? In the last section, we described the fact that aging processes appeared to slip past the bonds of natural selection. But we observe in our examination of appearances the presence of fundamental molecular deliberation. Are there really 'aging genes' that tell our cells to quit functioning in a certain way? Are their molecular signals which possess something that can actually measure the presence of time? Were these signals used for our benefit when we were younger, but when untethered from selection, work to our destruction when older?

Hang on. We've only just begun to explore the power of The Clock of Ages on our bodies. These questions will take on an even sharper focus when we get to our study of the aging of nerves. That is the subject of the next section.

6

The aging of the brain

It was a horrific painting. The focal point was a wild-eyed monster, inserting a bloody, headless corpse into its mouth.

'Are you sure this was created 200 years ago?' My colleague whispered in my ear. 'It almost looks like an abstract painting of a concentration camp. Or maybe a cartoon by Jeffrey Dahmer.' My colleague's comment, a reference to the late cannibalistic American serial killer, was with great amusement overheard by others in the auditorium. The comic relief was welcome, because of the grim nature of the subject – the probable deaths of famous painters. On the screen was a slide of a painting by Fransisco de Goya, created during one of his 'black periods.' I shuffled uneasily in my chair.

'The name is "Saturn Eating His Children," ' the lecturer began. 'It is an amazing example of Goya's artistic transformation, which began when he was middle-aged.'

The speaker described the fact that prior to his 46th birthday, Goya was a talented but absolutely conventional painter. His paintings were charming, picturesque, predictable, boring. But something happened to Goya that almost killed him. His brush with death unleashed a genius, perhaps a monster, and Goya would never paint the same way again. That something, and its cumulative effects on his art, was the subject of the lecture. I still found it difficult to watch.

'We don't really know what happened. We only know that it was sudden. In 1792, Goya became dizzy, giddy and nauseous, and started suffering hallucinations. He became paralyzed on his right side, acquired a ringing in his ears and his speech began to slur. Though he mostly recovered, the illness left him deaf and partially blind. It also left him depressive and very moody.'

The lecturer said that many explanations had been offered. In the mid-20th century, a British scientist offered a viral one. He had found a group of patients who had acquired an infection known as Vogt–Koyanagi disease. The patients suffering from this disorder, mediated by a virus, showed remarkably similar symptoms to Goya's. Some of the symptoms would subside; others, like partial blindness, would become permanent.

'But the most interesting explanation came in the early 70s,' the lecturer continued. 'A physician wrote a paper in the *New York State Journal of Medicine* stating that Goya's problem wasn't viral. It was environmental. He examined some unpublished material on Goya's painting habits, including the content of his paints. Basically, the physician concluded that Goya suffered from progressive and severe mercury and lead poisoning, due to the way he mixed his pigments.'

The role of heavy metal poisoning has been hypothesized for many medical phenomena of historical importance. But the article, in the lecturer's opinion, was not a faddish explanation. The physician showed that Goya suffered several bouts of these symptoms, beginning at age 32. Each time, the illness would keep him away from painting and he would recover. He would then go back to work, only to get sick again. The fact that Goya was the sire of 19 still-born or miscarried children gives further support to the hypothesis. Plumbism (the medical term for chronic lead poisoning) deforms human sperm.

'This poisoning, which is a savage attack on the central nervous system, in many ways mimics the symptoms of old age,' the lecturer continued. 'But with a difference. After the age of 46, Goya began to paint with ruthless genius. His dark change in personality, affected perhaps by his physical disability, launched him into a career that made him more than a passing footnote to history. In fact, it placed him into a trajectory that some consider timeless. One may only look at this piece . . .' the lecturer gestured to the slide, '. . . to see it.'

In the end, the lecturer summarized, the brain that gave Goya so many immortal moments also killed him. On April 16, 1828, he suffered a massive stroke and died. Whether he was a genius or not, I spent the rest of the day trying to erase the horrifying images seen in the slide show. I imagine Goya would have been pleased.

What we study next

In this section, we are going to use the career of Fransisco de Goya as an example of the aging process in the central nervous system. In many ways, the early experience of the painter's symptoms are similar to what occurs naturally in our later years. There is progressive degeneration of certain tissues that invariably leads to behavioral changes. To focus on this subject, this section will be divided into three parts. First, we will discuss some of the cognitive changes that occur as we get older. Then we will look under the hood of these changes and discuss their underlying anatomical substrates. Finally, we will examine the life spans of individual nerve cells, discussing several modern hypotheses that attempt to explain the changes in their function.

And we will begin with one of the greatest fears we have as the aging process enfolds our brains.

The changes

There's no doubt that one of the large concerns adults have as they get older is that they will grow 'senile.' The very word elicits a noxious perfume of prejudice and dysfunction: a terrifying mental incompetence, an embarrassing lack of control over bodily function, a recapitulation of childhood behavior, a reversing of the social role of offspring and sire. Many adults, when experiencing a mild brain-aging symptom (say temporary memory loss), become alarmed that the entire host of dysfunctionalities is about to be visited upon them.

All these fears have to do with a single aspect of the aging process: the effect of senescence on our nerves. One of the most obvious – and at the same time the hardest to pin down – is the effect on our thinking abilities. Hard scientific data are difficult to come by, because 'thinking' is so difficult to measure. As we age, there are types of thinking that seem to erode, and, at the same confusing time, other types that appear to sharpen. The literature outlines a few general trends and I will mention these below. What must be kept in mind at all times, however, is the subjective nature of the inquiries.

Intelligence

As there is no single definition of human intelligence, there is no adequate method of measuring an increase or declination as the years go by. Many professionals have used IQ tests in an effort to gauge some level of mental functioning. These were first developed in 1905 in an effort to distinguish retarded French children from non-retarded French children. Although they have some predictive power, many believe the examinations only assess a person's ability to take IQ tests.

The examination has been around for a long time, and, even accounting for certain changes, comparative scores can be calculated. A person's IQ peaks between the ages of 18 and 25. Thereafter, a slow decline occurs. What does that assess? Some researchers would claim that aging produces a reduction in the subset of intelligence factors the test is capable of measuring.

These data must be contrasted with other skills, however, skills that also affect IQ tests. A person possesses three times the vocabulary at age 45 that they do in their early 20s. From an overall perspective, the brain at age 60 possesses about four times the information that it does at age 20. Some researchers believe this expanded database allows us to make 'wise' judgements, decisions based on complex experiences and correlations that span decades. Still others claim that such mental acuity almost calls for another kind of social assessment, especially in the job market. Indeed, some lobby for the establishment of MQ (maturity quotient) tests. Such examinations, while filled with all the other problems of establishing IQ, would in their opinion begin to redress the age bias.

Memory

But is every measurement of aging in human brains a matter of subjectivity? The answer is no. The brain undergoes physical changes as we get older. It shrinks in size and becomes more filled with fluid. It even loses weight, typically about 5–10% from ages 20–90. Such changes are thought to do *something*, although exactly what, remains mysterious. Many researchers have examined the brain's ability to remember things, and found there to be a correlation between age and this cognitive ability.

To understand the research, one has to recall that there are several kinds of memory associated with human brains. There is declarative memory, a talent that allows us to remember things like 'Don't park your car under a tree full of birds.' Another kind of memory is procedural memory. This is the memory of activity, one allowing you to ride a bicycle, even if you haven't seen one in years. There is short-term memory, a kind of air traffic control tower that keeps conscious track of certain new thoughts (for a short period of time) as they enter your conscousness. And there is long-term memory, that allows you to recognize Aunt Jane, even if she hasn't been to your house in a long time.

The hardest part about memory research is trying to relate all these categories to biological substrates. And, in aging research, correlating specific behavioral changes with actual physical modifications in distinct regions of the brain. The reason is that those individual memory systems discussed in the previous paragraph engage many thousands of nerves in multiple regions of the brain. Defining how aging processes interact with these systems and thereby account for the observed cognitive behavioral changes is a significant challenge.

The bulk of the evidence suggests that age-related changes occur in two regions of the brain. The first region is the frontal lobe, an area of the brain immediately behind the forehead and eyes. It is felt that certain executive memory processes supported by this region degenerate with age. The other region affected by age is a C-shaped area deep inside the brain. This structure is known as the hippocampus. There is dramatic deterioration of explicit memory capacities supported by this region as we get older.

The effects of these changes show up in many ways in the research laboratory, even if correlation with a specific region is sometimes only a guess. For example, it is generally true that most humans exhibit memory loss by age 30. These deficits normally can't be detected unless they are tested, however. For example, a test to measure short-term memory is the examination, and expectation of recall, of a list of 24 words. When this test is given to a 20-year-old, he or she typically remembers 14 of those 24 words after a specified period of time. Under the same conditions, a 40-year-old person can recall only 11. A 60-year-old can recall only nine and a 70-year-old, only seven. This is one example of a large body of data that convinced many researchers that short-term memory changes as we get older.

There is evidence that long-term memory is affected as well. But it appears that most of the problem is caused not by an irreversible loss of a specific fact, but from impaired retrieval mechanisms. This can be demonstrated in a fairly dramatic fashion. A surgeon can stimulate an open brain with electrodes. Because there are no 'pain sensing' neurons in the cerebral tissue, the patient can be kept awake on the operating table. As different areas of the brain are touched with the electrode, specific memories are stimulated. Many of these are events or places that have not been recalled to conscious thought in years. First demonstrated by the pioneering work of Wilder Penfield, the location of these long-lost memories are so specific that a 'memory map' of a person's brain can be constructed.

As a result of these and other data, many researchers are focusing on how the aging process affects our ability to recall the memories stored within our brain. The evidence is mounting that the retrieval mechanisms, rather than the actual memories, are degraded as the aging process envelops our brains. Such selective impairment may be the source of a phenomenon experienced by many an aging human: having an intuitive understanding of a fact, but not being able to recall exactly what that fact is.

Sleep

One way to measure age-related neural changes is to examine functions that humans must undergo that also affect human thinking. One such function is our ability to sleep. Like aging, the reason why sleep is mandatory is not really known. The most obvious answer is that some form of recharging occurs. As we'll see below, however, there is lots of electrical activity in the brain during sleep and a fair amount of body movement. Thus, the restorative function of sleep is not solely to conserve energy. We mostly comprehend the essentiality of sleep by observing what happens when we do without it. It is necessary for human happiness and function, and when deprived of it, serious cognitive changes occur.

The amount of sleep we experience is not a hard-wired constant throughout our life. When we were toddlers, we slept as much as 20 hours a day. By the time we reached our first birthday, that number had shrunk to 13 hours. In our teen years, we get only about nine hours of sleep. By the age of 40, that number has been reduced to seven hours of sleep, six at age 50 and five hours after the age of 65. This does not mean we *need* fewer hours of sleep. It only means that on average, we *get* fewer hours in which to snooze.

From these numbers it is easy to see that sleep patterns change during aging, a fact that may have broad consequences for the quality of life in later years. In order to comprehend these alterations, we must first know the patterns that exists during a normal sleep cycle. We will discuss a typical eight hour sleep in a 20-year-old woman.

With the advent of increasingly sensitive EEGs (recall that these machines can detect brain waves), sleep researchers have detected the presence of unique patterns of electrical activity during a night's repose. These can be divided into overall categories, called slow-wave sleep and paradoxical sleep. These levels of arousal are mediated by a part of the brain named the reticular formation.

Slow-wave sleep can be further subdivided into four stages:

Stage 1 A period of drowsiness, when you are first going to sleep. The EEG is rapid and desynchronized, showing the patterns more of normal consciousness than of normal sleep.

Stages 2 and 3 Deeper sleep. The EEG has begun to slow down considerably. The waves are also increasingly synchronized, showing regular and predictable patterns.

Stage 4 Deepest sleep. This stage is mostly characterized by its distinctive waveforms. The pattern on the EEG shows large and peaked brain waves. This the period when we are most 'shut down'. No dreaming takes place and most body functions have low levels of activity.

Age-related changes

There are fairly predictable changes in these sleeping patterns as we get older. The first has to do with stage 1 of slow-wave sleep. As the years go by, there is an increase in the amount of time we spend lying awake at night. While the number of minutes we sleep remains fairly constant, it takes us longer to get into our slow-wave regimen. And if we get up at the same time every morning, we will suffer an increasing 'sleep debt' as we age. This decrease in overall sleep begins at about age 30 for men and about age 50 for women.

As we leave our 20s, we undergo increasing periods of wakefulness during the night. These periods capitulate the entire sleep cycle, the EEGs showing a fully aroused state. In later adult years, as much as 20% of the night consists of these fully aroused periods. By the age of 65, an unbroken night of sleep begins to become a rare phenomenon. Research on people between the ages of 73 and 92 showed that an average of 21 awakenings occurred during a given night. Men experience these more than women until about the age of 70. After that age, both genders experience the same patterns of interruption.

The Clock of Ages dictates a steady loss in our ability to maintain a tranquil night's sleep. The sleep in older humans is less restful, less deep and punctuated by fitful, wakeful consciousness. As we age, it is even easier to rouse us, regardless of the stage we experience.

The most interesting aspect of trying to understand age-related changes in intelligence, memory and sleep patterns is that our ability to experience them may be linked together. Aging may act on such a common linkage in order to produce its varied affects. One of the most interesting examples of this linkage occurred to a young American war hero in 1960 and is described below.

Staff Sargent Willis Boshears, a decorated Korean war veteran, was stationed at the US Air Force base in Wethersfield, England. He was out celebrating the New Year and invited a female acquaintance named Jean Constable (along with her date) to his house. Sgt Boshears soon passed out from his drinking. Later that night, Ms Constable's date, after briefly rousing Boshears to tell him he was leaving, went out the door. Sleepily, Boshears acknowleged the man's departure and then went back to his slumber. The next thing Boshears remembers is that he was aware of the woman's body beneath him, with his hands around her throat. Not until he was fully awake did he realize that the woman was Jean Constable and that he had just killed her. He was, of course, horrified by his deed, was arrested and eventually stood trial.

The 'sleepwalking' murder of Jean Constable represents a classic connection between several of the cognitive processes we have just discussed: awareness, sleep and memory. Sgt Boshears had murdered someone while he was still sleeping, all the while experiencing different levels of awareness. And he was not the first human recognized to have committed a crime while sleeping. Western legal tradition acknowledges the linkage between sleep and memory and awareness, and, in fact, Sgt Boshears was acquitted of all charges. But recognition does not mean understanding. Exactly how memory can be shut off, yet intelligence given full throttle in the midst of a sleep response is a prime example of just how hard it is to understand the mysterious partitioning within human brains. The fact that scientists who study aging must also confront this partitioning is one of the reasons this research is so hard to do. And why the data seem vague and, at times, almost anecdotal.

Even if the partitioning of our brains were fully understood and the behavioral connections between various cognitive talents known, the data would be intrinsically unsatisfying. The results describing changes in our intelligence, memory and ability to sleep yield only a cursory description of 'what' happens. They leave out almost entirely the all-important question of 'why?' Or better, 'how'? In order to understand this behavior at a more exacting level, we would have to open up the brain and begin examining individual nerves. Many researchers have done exactly that, and have attempted to describe some of this behavior in terms of cell groupings and individual nerves. The results of their almost heroic efforts are described next.

Neural reasons

The events occurring under our skull cap are of course mediated by nerve cells, the archetypal neurons. A typical human brain is composed of 100 billion of these cells, all wired together in a massive tangle, containing about 100 trillion different connections. Myriads of chemical reactions occur every second between these connections. All are highly coordinated, all unbelievably efficient, all very energy intensive. Indeed, 20% of the food you consume is shunted off to the brain for its varous tasks. Intense cerebral activity can burn as many calories as muscles during exercise.

A single neuron looks something like a scared mop. The handle end of the mop is called the axon, the shaggy strings on the other end are called dendrites, and the place where the strings meet the handle is called the cell body. Even though there are many types of neuron, they all share this same basic architecture. And they all share the same mission, which is to talk to each other with the frequency of teenagers on a telephone.

The scientist who discovered many of these structures was a turn-of-the-century researcher named Ramon y Cajal. Previous to his work, it was thought that the brain was composed of something equivalent to oatmeal – essentially an amorphous goo that somehow achieved complex cognitive tasks. But when Cajal used a then-newly developed cellular stain on a slice of brain, he didn't see globby porridge, he saw delicate lace. The apparent individual cells looked like finely crafted crystal, or tiny sun-like objects – and he gave the neurons he observed names like 'chandelier' cells and 'stellate' (star) cells. As with most anatomists of his day, Cajal drew pictures of what he saw, fine gossamer-like sketches of this most complex world. He would have no idea how amazing this world would be.

The beautiful branches Cajal saw had more than just esthetic appeal. Their structural morphology served a communicative function (Figure 17). As you may know, the language of nerves, like telecommunications, is electricity. When someone places an EEG on your scalp, they're really placing a crude wiretap into the conversations occurring in your head. Every time you raise your arm, blink your eyes or tap your toe, hundreds of nerves chatter and squawk to each other in an amazing electrical dialect. As you read this book, thousands of nerves in the back of your brain are busy assimilating and communicating to the neighbors what you see.

And exactly what are they saying? Nerves talk to each other by sending the chemical equivalent of tiny lightening bolts down each other's axons. When one nerve receives a stimulus it 'fires' a charge all the way down to its broom-handled end. It's an enthusiastic, if finicky event. If the stimulus is not of sufficient strength, the neuron will just sit there on its cellular haunches. But if the nerve does get enough of a stimulus, even if the impulse just barely makes the grade, the nerve will go off like a firecracker on a holiday.

Once the impulse is received, it travels in a specific direction: from the 'mop' end (dendrites) to the 'handle' end (tip of the axon). This it does at about 200 miles per hour. Once it gets to the tip of the axon, the neuron's next job is to try to get its neighbor as excited about the input as it was. This is not as easy as it sounds. The neighbor has placed a boundary, about a millionth of an inch wide, between it and the cell that just got excited. This boundary, called a synapse, must be overcome in order to pass along the signal.

The illustration on page 123 shows one such synapse. Don't be fooled by its apparent singularity, however. There are cells which form 100 000 other synapses with their neighbors. I am reminded of this when I consider how my university office used to be set up, two people in it, each with their own phone line. There were times when both phones would ring simultaneously. I would take up one, ask the party to wait a minute, and then pick up the other. Once both phone calls were for me and I distinctly remember getting slightly confused and having to say to one of them 'I'll call you back.' When I behold the talent that lies deep within my skull, I realize some humiliation: there are nerves that can handle both of those conversations and coordinate 99 998 other conversations as well!

HOW NERVES TALK TO EACH OTHER

Nerve cells are not physically 'connected' to each other. To communicate 'information' from one nerve to the next, a gap (termed a synapse) must be crossed. The nerve possessing the 'information' is called the pre-synaptic nerve. The neuron about to receive is called the post-synaptic nerve. (Two such nerves are shown in the box to the right.) The pre-synaptic nerve communicates its information with the use of gap-crossing neurotransmitters. Shown below are the chemical steps occurring between nerves to facilitate information transfer.

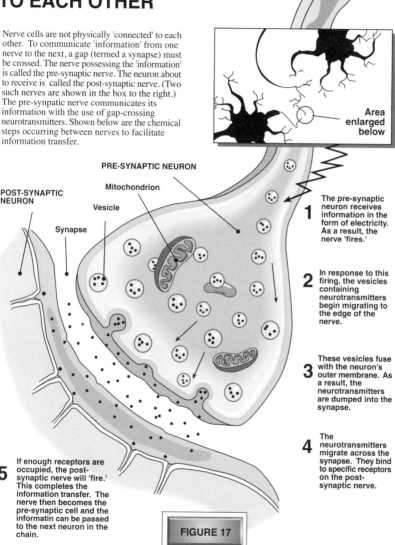

Area enlarged below

PRE-SYNAPTIC NEURON

POST-SYNAPTIC NEURON

Mitochondrion

Vesicle

Synapse

1 The pre-synaptic neuron receives information in the form of electricity. As a result, the nerve 'fires.'

2 In response to this firing, the vesicles containing neurotransmitters begin migrating to the edge of the nerve.

3 These vesicles fuse with the neuron's outer membrane. As a result, the neurotransmitters are dumped into the synapse.

4 The neurotransmitters migrate across the synapse. They bind to specific receptors on the post-synaptic nerve.

5 If enough receptors are occupied, the post-synaptic nerve will 'fire.' This completes the information transfer. The nerve then becomes the pre-synaptic cell and the informatin can be passed to the next neuron in the chain.

FIGURE 17

In order to perform this miracle of coordination, the nerve has to transmit its excitement across a physical space. How does it do that? When examined closely, the tip of the nerve has the molecular equivalent of a tiny navy, complete with cargo ships. This navy is just waiting to receive the signal to leave the cell, cross the space, and dock in its neighbor's harbor.

This leaving and docking is exactly what happens in a nerve cell. Only the navy isn't made out of ships, it is made out of a set of molecules known as neurotransmitters. When the signal reaches the end of one nerve, the navy is 'called out' and the neurotransmitters are dumped into the synapse. They sail across the gap (at a millionth of an inch, it's not a particularly wide chasm), and land in the neighbor's harbor. Only this harbor is not made out of docks and cargo platforms, it's made out of receptors on the cell surface. If enough neurotransmitters bind to enough receptors, the neighbor will get the message. In response, it will fire its neural lightening bolt and the whole process will repeat itself.

This is what happens, millions of times a second, whenever your brain tells your body to do something. The picture is much more complex, of course; there are other cells besides neurons lurking in its twists and folds, for example. These cells, called glial (for glue) cells, work like intelligent scaffolding, about ten glials for every one neuron. They serve much more than structural support functions – they can even 'fire' – but exactly what role they play in human thinking remains something of a mystery.

The Clock of Ages has a great deal to say about the aging of neurons and glial cells. It affects the ability of a nerve to fire, the way individual neurons relate to each other and, at some level, specific cognitive abilities. Exactly how some of this occurs is described next.

The synapse

Since the main job of nerves is communication, our ability to understand how nerves age must focus on the structures and molecules which allow nerves to talk to each other. Researchers have asked questions like: are there any changes in aging nerves one can observe in the synapse (that gulf between nerves we discussed)? How about neurotransmitters, those chemical messengers an excited nerve sends out over the gap? What about the docking molecules on the receiving nerve?

The answer to those questions appears to be 'yes.' The aging patterns are unevenly distributed, and different regions of nerves actually suffer different amounts of damage. As we get older, for example, there is a reduction in the amount of neurotransmitters in specific motor-controlling nerves. This is a specific destruction; a similar reduction is not seen in neighboring cells. The nerves in the regions that control sleep alter their neurotransmitter content with age, but once again adjoining cells are left alone.

There are examples of cells that lose their ability to receive signals from fully functional neighbors. The reason? Receptor molecules that would normally bind to incoming neurotransmitters are missing. Curiously, this loss occurs in regions also known to be associated with Parkinson's disease.

A number of neurotransmitter 'pathways' (a series of nerves that utilize and are biochemically influenced by the same neurotransmitter) appear to be affected as we age. With the advent of drugs that can alter these pathways in humans of any age, some interesting experiments on nerve aging have been performed. For example, a specific pathway known as the 'cholinergic' pathway has been shown to decrease in elderly people. This event is associated with memory loss. Young adults have been given drugs that artificially change this pathway, temporarily 'aging' their brains. When their memories are tested, these young people score exactly the same as older people who have not been given the drug.

These changes point to the fact that the aging process can produce subtle, almost delicate operations in our nerves. We don't know much about these 'quiet' changes, mostly because the technology needed to find all of them is mostly unknown or untested. This state of ignorance is changing, especially with the advent of the techniques of the molecular biologist (called in the popular press genetic engineering). I'll have more to say about this progress in Chapter 14.

There are other changes our nerves experience as they age which are not so subtle at all. These more dramatic alterations have been known for a long time and are described next.

The overall characteristics

The underlying process that describes how human nerves age can be summed up in two words: nerves die. In fact, it's a massacre. You lose between 30 000 and 50 000 nerves a day. By the age of 65, almost one-tenth of the brain cells you had as a young adult have vanished. All the synaptic connections disappear as well, and unless there is redundancy, so will the behaviors they mediate. The tragic aspect of this loss is that there is no regeneration. No new neurons can form once you pass what is called the perinatal stage of development (about four weeks after birth). Because these nerves don't regenerate, the loss will be permanent.

The incredible erosion of nerve cells has spawned two theories concerning the aging of human nerves. One theory is the depressing Neuronal Fallout Model; the other is the more hopeful Neuronal Plasticity Model. We will briefly examine salient points in each theory.

Neuronal Fallout Model

Considering the number of nerves that exist vs the number of nerves that die, a large question is: how many nerves have to die (and *where* must the destruction occur) before a particular behavior is extinguished? This question isn't easy to answer. The brain has this nasty habit of representing the same information in multiple sites. There is evidence that even the characteristics of a single object are parceled out to various physical regions in the brain. Recall, for example, the patient who couldn't recognize a photo of a rhinoceros but could describe and draw one if the letters r-h-i-n-o-c-e-r-o-s were presented.

Even though such redundancy occurs, the Neuronal Fallout Model assumes that the attrition will eventually show up in behavioral loss. This model infers that, as we age, progressive neural loss will exhaust the brain's reserve capacities. This destruction results in the decline of sensory, motor and integrative functions. If a particular 'master' nerve which may control the functional status of many other subordinate nerves dies, a cascade of degeneration will occur. All of the functions it mediated will be lost. Considering the enormous interactive neural webs within the brain, this loss might reverberate through distant regions, changing patterns of communication system-wide. This depressing model posits inexorable, unflinching and deleterious behavioral effects due to the absence of nerves in aging people.

This explanation of central nervous destruction is not the only theory that explains the aging of our nerves and brains. Another set of hypotheses, ones that are a little more optimistic, have also been proposed. To describe them, I would like to illustrate a fascinating flexibility in nerves that results in an interesting talent. And to do that, I would like to talk about a man who lived in the 18th century, one Jedediah Buxton.

Jedediah answered the question with a 28 digit number. The journalists in the room gasped slightly and then started scribbling furiously on their notepads. He had just been asked the following question: 'How many eighths of an inch exist in a body whose three sides are 23 145 789 yards, 5 642 732 yards and 54 965 yards?'

A 40-something Jedediah Buxon, calculated to have a mental age of ten, stared languidly out the window and then blurted out the correct number. His answer was cross-checked by a physician who happened to be in attendance. The journalists scurried out the door, information in hand, to meet their deadlines. Jedediah, left with his attendants, asked for one of his toys and walked impatiently around the room. Just another dumb game his adult friends wanted him to play. His talent has mystified researchers for centuries.

Science has documented a number of people with challenged intelligence who can perform miraculous intellectual and artistic feats. In crueler times, such people were called idiot savants. Although such talents give us great clues into the mysteries of brain development, most of us are not quite sure which mysteries are being addressed. It is known that damage suffered in the left hemisphere of our brain can be compensated for with increased neural activity in the right. Some researchers believe this can 'overdevelop' and grant supernumerary capabilities to injured gray matter. There have been hypotheses put forth seeking to explain behavior like Jedediah's through such neural plasticity. It may be that the growing fetus 'knows' that something is starting to go wrong with its brain development. In an effort to compensate, a sudden burst of neural activity, recruiting other nerves and establishing new connections, may make certain regions 'smart'. The result? A retarded child with unbelievable talent.

Whereas this kind of flexible programming is still very much of a hypothesis, the fact of neural plasticity and changes due to experience are solid fact. A theory has been developed around this flexibility that seeks to explain not only the development of our neural tissues, but also their aging. It is called the Neuronal Plasticity Model.

Neuronal Plasticity Model

Advocates of the Neuronal Plasticity Model have one word for those in the Fallout Camp. That word is: nonsense.

The heart of this alternative hypothesis has to do with the fact that neural pathways within the brain are not nearly so set in stone as was once believed. Neurons exist which grow new axons or longer dendrites when neighboring neurons are damaged. Even non-regenerating neurons may have the ability to change their synaptic associations with other healthy nerves. The biology of 'learning' may involve actual physical synaptic changes in the associations between existing neurons. These data do not point to some brain that slowly sets up like poured concrete, but a living organ fully interactive with its environment.

These ideas can affect deeply how the aging human brain is viewed. In the Neuronal Plasticity Model, the effects of the neural destruction need not be permanent. Other neurons can take the place of fallen neighbors, establishing new synaptic connections, preserving in part the 'lost' functions. Some researchers believe there is actually a net gain in the density of synapses as the years roll along. There is evidence that intellectual decline can be slowed or even reversed if certain cerebral 'training' regimens are employed later in life. New pathways may be constructed simply by undergoing new experiences or new training. In this model, the brain is not treated like a deteriorating, progressively crippled old organ. Rather, it is considered to be an active, resilient thinking machine, fully capable of interaction regardless of the comment of The Clock of Ages.

The resolution

Is there a proper resolution to these two ideas? The answer is 'not yet.' There is still so much to learn about how neural connections affect outward behavior, that most of what we think happens must fall into the realm of 'comment.' For example, many have pointed out that older citizens continue to make tremendous contributions well into their old age, and that this fits in quite nicely with the plasticity model. Bertrand Russell wrote *Human Society in Ethics and Politics* at the age of 82. Leopold Stokowski conducted the debut of Charles Ive's Fourth Symphony at the age of 83. Frank Lloyd Wright began drawing the architectural plans for the Guggenheim Museum at the age of 88. He would live to see it fully constructed, celebrating his 90th birthday in the process. The idea that human minds must inevitably shut down in their maturity should take a back seat when confronted with these accomplishments. Or at least, we'd like to think so.

Others point out that we remember these achievements precisely because they are rare. They point to the legions of patients in adult care facilities with crippling dementias, the number that suffer Alzheimer's disease, those older people with severe memory loss, with severe strokes, and so on. These advocate the idea that because neural loss is inexorable, severe behavioral changes are inevitable too. And this, regardless of the plasticity of the human brain.

Whether one believes in Neuronal Fallout or Plasticity, the central biological fact of neural cell loss must be confronted. The question is whether the observed plasticity can completely overtake or compensate for the losses observed in aging human brains. For that question, the jury is still out. In fact, the jury may be confused. Since we are not yet at a point where we can ask a nerve what its favorite color is, we don't know whether Bertrand Russell was the norm or the exception. We can only hope that the data we discover will match our hopes and expectations.

In the next section, we are going to take an inventory on this cellular loss. We will go through the brain region by region, first describing the function, and then, the effect of aging. I must caution against a certain temptation, however. Because we know a lot about the job descriptions for nerves within these locations, it may be appealing to speculate that neural loss means functional loss. Remember that when we discuss these regions, we are talking about evolution's trickiest brain. This balance between loss and plasticity must ever be kept in mind to get an accurate picture of how aging affects us mentally.

Of dinosaurs and neurons

While cataloging neuronal loss in the brain, we have to know something about its structure. For that we will take a quick tour. Digging into the vast complexity of the human brain reminds me less of a neuroanatomical lesson than it does of a fossil dig. Before we begin our excursion into the world beneath the cranium, let me briefly explain what I mean.

As a little boy, I was always fascinated by archeological digs – especially if they contained old dinosaur bones. Knowing this tendency in her little boy, my mother once fashioned some 'ancient' dinosaur skeletal remains out of a couple of old steak bones. Then she buried them in the ground and the next day (my birthday), told me to go looking. To my great delight, I found her creative attempt at a reptilian vertebrate. We spent the rest of that morning happily constructing our own prehistoric bones and reading old dinosaur books together.

One of the great tools archeologists use to characterize the age of an artifact in a site has to do with the depth at which it was discovered. Stated simply: the deeper you go, the older things usually are. Objects and structures discovered on the surface are much younger than those discovered deep below. This method has been used to gauge relative ages of artifacts in a single dig. It has even been used to create hypotheses about how dinosaurs lived and when they became extinct.

This same principle of deeper levels uncovering ancient structures has as much to do with the anatomy of the brain as with an archeological site. The overall plan of our cerebral architecture appears to possess this same 'the-deeper-you-go-the-older-you-are' quality. There are ancient structures in our skulls that we share with most vertebrates; middle-aged structures that we share with only other mammals; and brand new structures that we share with no other creature on earth. We will begin our tour by examining some of our oldest structures (Figure 18).

The brain stem

Most of us have an intuitive understanding of what the brain looks like. Its two hemispheres appear to be stuffed into our skulls like twin pythons, with massive twists and turns, hills and valleys found throughout its architecture. These structures appear to be wrapped around a single twig – our spinal column – which enters our brains from below. This part of the column is called the brain stem.

The brain stem has many ancient functions, all shared by other terrestrial vertebrates. It also controls our breathing and regulates our heart beat. The brain stem contains the reticular activating system, which, as you recall from our discussion on sleep, regulates our resting and waking. It also contains an elongated group of nerves called – get ready for a big word here – the locus coeruleus. These cells are connected to areas much higher in the brain. If something gets us excited or alarmed, these cells create the neural equivalent of a signal flare, alerting our higher centers to the event. This locus may also have a lot to do with waking from sleep, interacting with certain neurotransmitters to mediate arousal – and possibly even dreaming.

The brain stem is one of those structures that loses very few neurons with age. Only one place, the locus coeruleus, suffers damage. This occurs suddenly, around the age of 65, with almost 40% of the nerve cells dying. Cell death in this area may be one reason why it is increasingly difficult to get to rest as we get older. But, as mentioned previously, this idea lies only in the realm of speculation.

The cerebellum

Moving slightly up and sideways along the brain, we come to a structure termed the cerebellum. In dissection it looks something like broccoli. This shrivelled mass of cells has many functions, including the ability to process and control our muscles, joints and tendons. The emphasis is on the word control. The fact that we can move in a coordinated fashion, rise to a given posture and not fall down when we walk or run all has to do with the switching mechanisms within the cerebellum. These integrative functions are mediated by some of the most complex cells in our brain, called Purkinje fibers. One cell alone can crosstalk with tens of thousands of other nerves, talking to the higher centers of our brain, coordinating our movements.

There is some evidence that the cerebellum defines boundaries for our motion. When we reach down to tie our shoe, for example, our legs don't start jogging and our arms start reaching for a cookie jar. This restriction to our hands touching our shoes and our legs bending to accommodate may be due to the editorial control of the cerebellum.

The cerebellum is a structure that loses a lot of cells in our later years. Particular attention has been paid to the Purkinje fibers. As you recall, these complex neurons communicate with myriads of other nerves. They coordinate much of our motor functions, receiving commands from other regions and then executing the instructions.

During the aging process, those Purkinje fibers begin to die, with the largest loss occurring suddenly after the age of 60. During our life span, we will typically lose 25% of what we had as a young adult. This loss may have a direct effect on our motor function. We may not be able to move as quickly, maintain our balance or correct posture as we get older. It may get increasingly difficult to issue commands that our arms and legs understand as coordinated movement. Fine movements in rapid succession are particularly vulnerable. If neural commands aren't issued to our muscles, as you recall, they will begin to atrophy. All of this lack of

How the brain is organized

*To understand the brain's aging process, it's important to know some neural
anatomy. Here's a quick overview.*

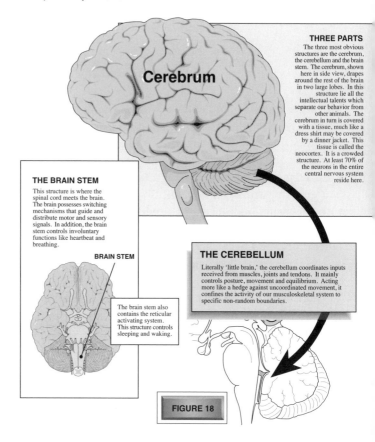

Cerebrum

THREE PARTS
The three most obvious
structures are the cerebrum,
the cerebellum and the brain
stem. The cerebrum, shown
here in side view, drapes
around the rest of the brain
in two large lobes. In this
structure lie all the
intellectual talents which
separate our behavior from
other animals. The
cerebrum in turn is covered
with a tissue, much like a
dress shirt may be covered
by a dinner jacket. This
tissue is called the
neocortex. It is a crowded
structure. At least 70% of
the neurons in the entire
central nervous system
reside here.

THE BRAIN STEM
This structure is where the
spinal cord meets the brain.
The brain possesses switching
mechanisms that guide and
distribute motor and sensory
signals. In addition, the brain
stem controls involuntary
functions like heartbeat and
breathing.

BRAIN STEM

The brain stem also
contains the reticular
activating system.
This structure controls
sleeping and waking.

THE CEREBELLUM
Literally 'little brain,' the cerebellum coordinates inputs
received from muscles, joints and tendons. It mainly
controls posture, movement and equilibrium. Acting
more like a hedge against uncoordinated movement, it
confines the activity of our musculoskeletal system to
specific non-random boundaries.

FIGURE 18

THE DIENCEPHALON
(brain shown in cross section below)

THE LIMBIC SYSTEM

The limbic system is a combination of structures projecting deep into the brain. Collectively, they control such emotional responses as rage, joy and fear.

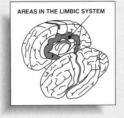

AREAS IN THE LIMBIC SYSTEM

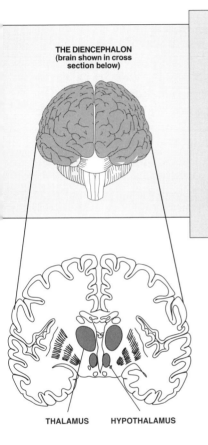

THALAMUS   HYPOTHALAMUS

The diencephalon contains two important structures, the thalamus and the hypothalamus. Part of the thalamus' job is to regulate and distribute sensory information (signals received in the eyes, ears, skin, etc.) to the rest of the brain. The hypothalamus controls, among other things, body temperature, blood pressure, appetite and circadian rhythms (sleep/wake synchrony associated with light).

coordination serves to restrict our range of motion as we age. A large part of this may be mediated by the loss of cells deep within our cerebellum.

The diencephalon

As we move away from the cerebellum and up and out of the brain stem, we encounter the brain's most inner layer. This region is called the diencephalon. There are two dominant neural landlords in this area. One landlord is called the hypothalamus. It is probably the world's smallest diplomat (not much larger than a thumbnail), constantly negotiating between the brain and the body. Emotions speak to us through our hypothalamus. This organ regulates our body temperature and our blood pressure. It even controls our appetite and if externally stimulated, can send us into rage and fear. It is in constant communication with our pituitary gland, the seat of hormone regulation, and another structure, which you can think of as Big Ben. Or, perhaps, Little Ben. This tiny organ is called the pineal gland, and contains an internal, light-sensitive 'clock.'

Directly on top of the hypothalamus is the other landlord, termed the thalamus. Like the cerebellum, this organ's major function is input control, not from our muscles, however, but rather from our senses. The fact that you are reading this page right now is because your thalamus is busy receiving signals from the eyes and shuttling those inputs into the proper areas of the brain. The thalamus also controls our hearing, our sense of smell and any other input from a sensory organ. Without it, our brains would be literally incapable of interacting with the rest of the world.

The Clock of Ages has a very different effect on the aging of the diencephalon when compared to the aging of the cerebellum. The number of nerves and their infinite connections look the same at 60 as they do at 20. This is true whether one examines the connections between the hypothalamus and the pituitary, the nerves deep within the thalamus or the fibers projecting from the pineal gland. These untouched areas serve as an example of how selectively the aging process sweeps over the brain. Why it leaves these structures alone yet deeply affects neighboring structures millionths of an inch away is one of The Clock of Ages' great mysteries.

The limbic system and the hippocampus

As we move outward from the territory of these old landlords, we come to a region that used to be called the 'smell brain.' The reason? The areas mediating our olfactory abilities sit on the top of the diencephalon. It turns out that this system, known as the limbic system, is very important to our emotional well being. And, believe it or not, to our ability to learn things. In animals, you can stick a needle-thin electrode into the limbic system and do some pretty interesting tricks. If you apply a current to one area, the animal will suddenly leap into rage. Stimulate another area, and fear is created; yet another area, and joy will be triggered. This has even been done a few times in humans, demonstrating that we share its function (as well as its structure) with most vertebrates. It is truly an old part of the brain. These experiments demonstrated that a seat of the emotions really exists, right in a part of the brain we also share with lizards. It's not necessarily flattering.

The limbic system is also associated with an area known as the hippocampus (literally, seahorse). This organ has been linked with our ability to transfer events residing in our short-term memories to our long-term memories. The link between our memories and our emotional experiences may not just be psychological, but in some fashion also neurological.

These areas of the brain, especially the hippocampus, are somehow one of aging's favorite targets. Starting at about age 30, various regions of the hippocampus start to die. The loss is gradual and by the time we reach old age, about 30% of the neurons in this area are gone. Is this the reason why so many of us experience memory loss, or find difficulty in learning, as we get older? No one really knows the answer to that question, although the implications are both tempting and, to some researchers, obvious.

The cerebrum

The limbic system is surrounded by the giant, mushroom-like structure of our cerebrum. It is divided into two halves, or lobes, and these in turn are subdivided into specific regions. The cerebrum is itself covered with a kind of neural wrapping paper called the cortex, which is about one-hundredth of an inch thick. Don't be fooled by its tiny width, however. Almost 70% of the neurons in the central nervous system are embedded in the cortex. If you were to unfold it like a picnic blanket, you would create a sheet one and a half feet squared.

This cortex is the newest part of the human brain, possessing cognitive machinery that separates us from the rest of the animal kingdom. It is within the cortex that we create symphonies, worry about the weather, covet our neighbor's automobile and wonder about the origins of life. The odd thing about its biology is that, unlike the lower parts of the brain, the cortex looks the same all over. Under the microscope it appears as a tangled mass of thorns. But it is hardly a tangle. Regions within the cortex are unbelievably specialized, creating and performing functions we are only just beginning to understand.

Is this complexity a target for selective neural destruction? The uncomfortable fact is that there is a dramatic loss in the number of neurons in certain regions of the cortex. Exactly what it means is an open question. What follows below is a list of the various structures and the percentage loss during aging in the cerebral cortex. Except where indicated, these losses compare numbers seen in human adults at ages 70–80 vs adults in their 20s.

Motor areas (frontal cortex)
The estimates of neuronal loss vary widely, depending upon the study. Between 20% and 50% of cells in these areas are lost with advancing age.

Visual areas (occipital cortex)
The cell density decreases by about 50% in these areas. A lot of this destruction occurs prior to age 40.

Auditory areas (temporal lobe)
The number of neurons decreases by 30–50% in these areas, depending once again on the study.

Prefrontal cortex (associated with various kinds of memory)
This area shows virtually no neuronal loss.

As we explore different senses, we will return to these areas and describe in more detail the effects of aging. Even though this loss might seem like a mixed bag of events (rendered even more confusing with data from multiple studies), there appears to be an overall pattern. The neurons that mediate our higher cortical functions seem to be untouched. Those nerves that affect our motor and sensory areas are more profoundly affected by the aging process (Figure 19). It's almost as if the aging process is saying 'I'm going to take away your animal behaviors, but I will leave intact the nerves that make you specially human.'

If The Clock of Ages possesses any such thing as biological mercy, this is probably its clearest example.

The strange case of Phineas Gage

I have attempted to limit the interpretation of neuronal death in the light of human behavior. While this is a responsible thing to do, I do not want to leave the obviously wrong impression that nerve damage and human behavior are unrelated. To illustrate this fact, consider an unfortunate and quite famous case first described in 1848. It had to do with the person-ality of a 25-year-old construction foreman named Phineas Gage.

By all accounts, Gage was a competent, level-headed guy. He was well liked by all his co-workers and had a reputation around town for being an intelligent, responsible businessman.

This all changed one morning in Vermont. Gage was working at a construction site when an explosion occurred. To the horrified amazement of the co-workers, the accident drove a large iron rod through Gage's head. As various onlookers rushed to his aid, he crumpled to the ground and lay still. Everyone thought he was dead. Everyone except Phineas Gage.

To the crowd's further shock, the foreman sat up and began speaking coherently. A local doctor was summoned, and the next day, the rod was removed. Everything seemed to be fine. Gage appeared healthy, had most of his motor skills, all of his memory, senses and so on. In fact, everything seemed 'normal' in the construction worker except one important charac-teristic: his personality. Over the weeks, it began to be apparent that something had dramatically changed with Gage's disposition. He became foul-mouthed and irresponsible. He waffled on decisions and soon lost his job. This irresponsibility would become permanent. Gage turned into a drifter who could not hold any kind of work. The rod had changed *something* in his brain after all. Exactly what, could not be ascertained, but the impact on Gage's life is an interesting testimony to how neural damage can affect human behavior. More than a century later, we still don't really know how.

Nerve cell survival in aging

We lose a lot of nerve cells with age. Here are a few survival statistics.

As we get older, specific structures within the brain suffer neural loss. Listed below are the percentage of cells remaining in a given area after age 60 (compared to the number of cells as a young adult). Note that some structures do not show any appreciable loss.

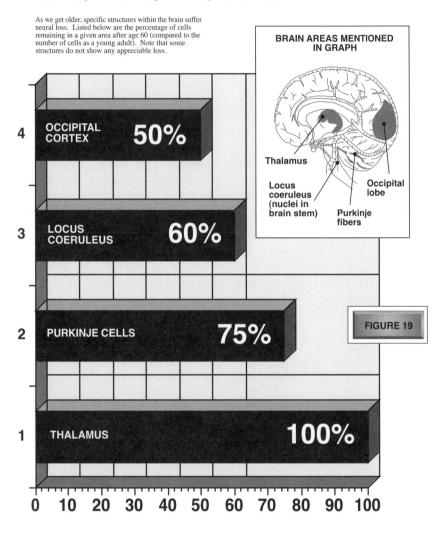

BRAIN AREAS MENTIONED IN GRAPH

Thalamus

Locus coeruleus (nuclei in brain stem)

Occipital lobe

Purkinje fibers

FIGURE 19

4 OCCIPITAL CORTEX **50%**

3 LOCUS COERULEUS **60%**

2 PURKINJE CELLS **75%**

1 THALAMUS **100%**

0 10 20 30 40 50 60 70 80 90 100

The tale of Phineas Gage brings up a point still relevant today. We are only just beginning to understand the association between particular nerves and behavior. We know that there is a daily loss of neurons that leads to a substantial change in human brain weight as the years pile up (in its young adult configuration, a human brain weighs about 3 pounds; by the time we reach 65, the brain will have lost more than an ounce of its overall weight).

In summary

Because there is selective destruction of nerve cells as we get older, parts of the Neuronal Fallout hypothesis are experimentally verified. Specific parts of the brain undergo unique and timed neural destruction. This mayhem can occur suddenly, with much of the activity happening in the opening years of our sixth decade of life.

But this destruction is practically the only data on which we can hang our hats. Why? Not only does neuronal loss vary from person to person, the very notion has to be restricted when one is discussing a living brain. Changes in neural architecture show great susceptibility to experience and environmental circumstance. Such changes can be observed not just externally but also internally. The number and distribution pattern of intracellular molecules (including neurotransmitters) is also influenced environmentally. Indeed, there is strong evidence that internal molecular changes lie at the heart of our ability to learn things at any age.

Considering such structural vulnerability and the overall number of nerves we possess, what does neuronal loss *really* mean? This question is the essence of the Neuronal Plasticity hypothesis, a notion that celebrates flexibility. It may mean that neural loss and neural agility work together. The plasticity may be a reaction to nerve death, a compensating series of reactions that allow us to function intellectually for longer.

Conclusions

All of this anatomy and chemistry can be intuitively grasped simply through experience. Even though there are great credibility questions linking specific neural losses to observable behavior deficits, we know that *something* changes as we age. These changes give us more vocabulary and less sleep. They allow us opportunities to share great wisdom, and a decreased ability to remember when we do.

They can even project us into whole new worlds. Some neural changes, as Phineas Gage's, result in a medical footnote that is more tragic than enlightening. Some neural changes, as Fransisco de Goya's, result in the dramatic revelation of genius. As we age, we enter our own new world, a place that allows us to view images through the lens of understanding.

Even if we sometimes find them difficult to watch.

7

How the heart ages

She was lying on the bed, her husband close beside her. He stroked her forehead as if it was china. 'Knowledge by suffering entereth,' she said, deliriously quoting her own poetry, 'And life is perfected by death.'

Elizabeth Barrett Browning, her husband cradling her body, was experiencing the last morning of her life. It was, by all accounts, an appropriate end to one of the world's great poets and to one of the world's more interesting love affairs. The illness that would take her life was no stranger; in fact most of Elizabeth's life was divided into repeating stanzas of accident and disease; it wasn't until she met her husband, Robert, that she received any respite from the various aches and pains life threw her way.

The place where Elizabeth suffered her inaugural trauma was also the center of her universe – her family home. One of 11 children, she grew up in a rich English household, dominated by an imperious father. She had been schooled there, learning to read and write in Greek, Latin, Italian, German and Hebrew.

Elizabeth fell from a horse at the age of 15. The spinal injuries she received made her a virtual shut-in. Her father, already repressive and overbearing, became the jailer in what was essentially a gentrified prison. She would be sentenced to the confines of her house for 16 years.

At the age of 31, Elizabeth suffered her next round of trauma. At first her family thought the poet was coming down with a cold. But she eventually began to spit up blood and to the confirmation of her worst fears, was diagnosed with tuberculosis. On the advice of the physician, Elizabeth's father now insisted on rest, or rather, solitary confinement. Her prison collapsed to the space of her bedroom. For five additional years, Elizabeth would be locked in her room, seeing very few people, save those her father allowed entrance. She became addicted to morphine, a common prescription at the time for tuberculosis. Elizabeth was allowed to write –

indeed, her publications and literary critiques reached a fairly wide audience. But were it not for the persistence of a promising young poet from the country, Elizabeth Barrett Browning might never have left her prison.

Robert Browning entered Elizabeth's life in 1841. His sunny disposition, incredible optimism and eventual love for this wan, bed-ridden woman transformed them both. They began to carry on a secret romance, his excuse for weekly visits being poetry readings and critique. In 1846 they were married, to the violent opposition of the elder Barrett. The lovers secretly left England a week after their wedding and eventually settled in Florence, Italy. Her father was so enraged he forbade mentioning the name of Elizabeth in his presence, vowing eternal wrath and unforgiveness. As you might expect, she never went back to England.

Elizabeth was to live for 15 more years. At first, she experienced an amazing recovery of health. Her lungs seemed to clear. Her energy returned and, with constant encouragement from her husband, wrote with vigor and insight. So changed was her physical state that one friend told her 'You are not improved, you are transformed!' She wrote of her experiences in England: 'I was buried (in) a morbid and desolate state . . . which I look back now to with a sort of horror.'

In 1849, at the age of 43, she gave birth to a healthy boy. But Elizabeth began to suffer miscarriages, and with a severe loss of blood, experienced profound anemia. She never recovered from her morphine addiction, though in Italy managed to control it with great resolve and force of will. Slowly but surely, Elizabeth's physical troubles returned.

The final bells sounded when she came back from a visit to Rome. As ever, the problem settled in her lungs. It was at first diagnosed as a cold, but later turned into a chronic bronchitis. She went back to bed and stayed, her husband by her side, attending to her physical needs. On the morning of June 29, 1861, Robert was ladling soup into a bowl beside her bed. Suddenly her expression looked vacant. He stopped what he was doing, and then blurted out 'Do you know who I am?' She smiled, kissed him and quoted her last words. And then, in his arms, she died.

A chapter devoted to hearts

The relationship between Elizabeth and Robert Browning remains one of history's great love affairs. He was to survive his wife by almost 28 years, but to the end felt his wife's death like an open wound. At the age of

77, Robert Browning suffered a massive heart attack and died shortly after. His wish was to be buried next to his wife at a graveyard in Florence. Alas, the cemetery was closed. Instead, he was buried at Westminister Abbey, a place where his remains still reside.

We are going to use the love of the Brownings to describe an affair with which Robert was very familiar – the effects of time on the human heart. We will not be nearly as romantic as the infatuated poets, though the lessons to be learned are no less interesting. The fact that the cardio-vascular system must feed all parts of the human experience – including the writing of verse – means that its aging is one of the most important subjects we will discuss.

To describe the effects of the years on our hearts and blood stream, this section will be divided into two parts. First, we will describe some of the relevant anatomy. Along the way, we will talk about the nourishing nature of blood, and how its oxygenation is critical to tissue survival. Second, we will talk about the aging process. We will describe how the blood's capacity to penetrate tissues affects human maturity.

The take-home lesson from this section is the central roles our hearts and blood vessels play in the senescence of all tissues. In a later section, we will examine how the performance of the cardiovasculature is amenable to external influences. In this section, however, we will discover how some parts of these systems can never be influenced. Like the effects of trauma on Elizabeth Barret Browning, we will discover the deterioration to be life-long and, eventually, life-extinguishing.

A cardiac introduction

Make a fist out of your hand and then squeeze it. That's about the size of a human heart undergoing a contraction. Relax your hand for a moment and there is your heart, now beginning to fill with blood. These two actions constitute its beating, which is broadcast throughout the body – like a radio wave – in the form of a pulse. The desire to contract is so strong that the heart will beat for a time if you remove it from the body. Or even if you dissect it. You can literally place tiny strips of the heart in a dish full of salt-water and the cadence will still go on.

The reason for this great cardiac determination has to do with its func-tion, which is to push blood into and out of the tissues of your body. The superhighway the heart uses is termed the vasculature, a fancy word for blood vessels. It is quite a highway. Almost 60 000 miles of vessels are crammed into a human being whose average height is less than six feet. And the speed limit is as fast as the highway is long. A typical blood cell leaving the heart, traveling to the big toe and arriving back at the heart takes about 12 seconds.

As you know, the human heart is divided into chambers. You can think of it as a tiny two-storey pump-house, with two rooms per storey (Figure 20). The upper rooms are called atria (the left and right atrium); the two lower rooms are called the ventricles. This pump-house is made out of cardiac muscle, which as you recall is one of the three types of muscle human beings possess. Although there are subdivisions, the main muscle is called myocardium. Like the skeletal muscle, there are tiny proteins within the cells of the myocardium that are responsible for the contractions.

How the pump works

The heart pumps blood because of a highly coordinated series of contractions between the floors and rooms of the 'pump-house.' First the upper chambers contract and then, on cue, the lower chambers do too. While one floor is being squeezed, the other floor is resting. That's how your heart 'sleeps,' taking tiny catnaps between contractions. These interactions are supervised by a 'mini-brain' stitched directly onto the heart's surface. This brain is a complex collection of nerves whose sole job is to make sure the heart is always in motion.

As you know, the purpose of this beating is to push the blood around – specifically to two general locations. First, the blood has to pick up oxygen, which means the heart has to push the blood cells into the lungs. Second, once laden with oxygen, the cells have to return to the heart for delivery to the rest of the body. This means they have to enter the heart twice – once to become loaded, and once for body-wide distribution. To understand exactly how it's done, we'll follow a blood cell in its unloaded form, getting ready to be pushed into the lungs.

An oxygen-depleted cell first arrives from its anatomical journey into the upper room of the pump-house – the right atrium. This atrium contracts, which pushes the blood cell into the room directly below the atrium, the right ventricle (the cell gets to the lower room by means of a trapdoor biologists call a valve). The blood cell sits there, waiting for the lower room – that right ventricle – to contract. Once it does, the blood cell leaves the lower room the same way a lucky mole might leave a suddenly flooded hole, by floating to the surface. In this case, the cell leaves the atrium by means of another trapdoor located in the room's ceiling. It is now outside the heart in a blood vessel, on its way to the lungs to pick up fresh oxygen. We'll have more to say about what happens there when we discuss the aging of the lungs.

After a visit to the lungs, the fully oxygenated blood cell has to get back to the heart. It does so via a blood vessel that dumps it into the other upper room of the pump-house, the left atrium. The atrium contracts, another trapdoor opens and the cell is now in the left lower room, the left ventricle. It is now almost ready to leave the heart and go find a tissue to feed. As you might expect, the left ventricle now squeezes, pushing the blood cell through one final trapdoor (also located in the ceiling). The cell now finds itself in the aorta, that familar, giant blood vessel connecting the heart to the rest of the vascular highway. From there the blood cell will migrate to a distant part of the body, unload its oxygen, and find its way back to the pump-house to start the whole thing over again.

There are many ways to measure the efficiency of this entire pumping action. One familiar term is the pulse rate, the number of beats per unit time. Another term is the stroke volume. This is the amount of blood pushed through the left ventricle (one of the lower rooms in the pump-house) and out the aorta for general distribution. There is the notion of cardiac output. This is the stroke volume multiplied by the heart rate. All of these functions are affected as our cardiac tissue gets older. Their measurement is a general reflection of our state of health.

Blood vessels

The heart and its ability to pump blood through our tissues is only one part of the aging story. The structure of our blood vessels is also affected as we get older. Before we discuss in detail the senescence of the cardiovascular system, we will need to briefly review some salient points of their anatomy.

Blood vessels can be divided into two categories. One type, the arteries, carry oxygenated blood from the heart to the tissues. As arteries get smaller, they are termed arterioles. When arterioles invade tissues, they

THE STRUCTURE AND FUNCTION OF THE HEART

Drawn below is the anatomy of the human heart, showing its four chambers. On the right is the progression of a heartbeat. This four-step process illustrates how the chambers work together to supply and distribute oxygen-laden blood cells to the body.

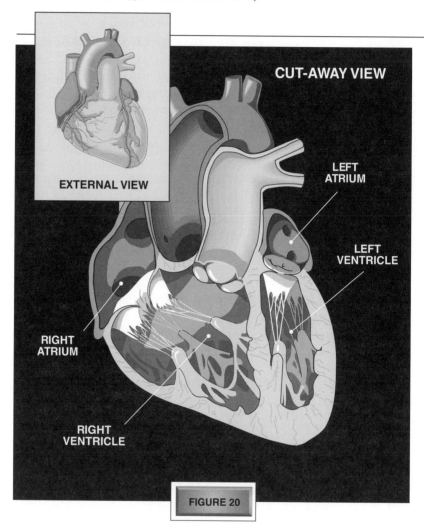

CUT-AWAY VIEW

EXTERNAL VIEW

LEFT
ATRIUM

LEFT
VENTRICLE

RIGHT
ATRIUM

RIGHT
VENTRICLE

FIGURE 20

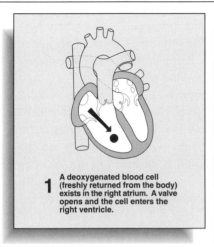

1 A deoxygenated blood cell (freshly returned from the body) exists in the right atrium. A valve opens and the cell enters the right ventricle.

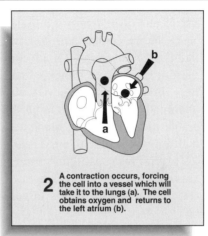

2 A contraction occurs, forcing the cell into a vessel which will take it to the lungs (a). The cell obtains oxygen and returns to the left atrium (b).

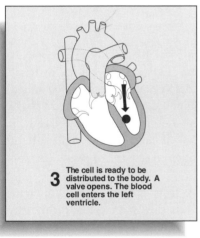

3 The cell is ready to be distributed to the body. A valve opens. The blood cell enters the left ventricle.

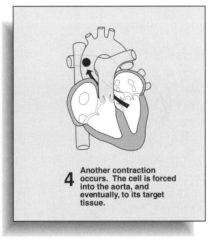

4 Another contraction occurs. The cell is forced into the aorta, and eventually, to its target tissue.

suddenly branch into countless tiny vessels called capillaries. It is through these thin walls that blood vessels unload their oxygen to various tissues.

An artery is essentially a hollow core surrounded by three concentric layers of tissue. The core is called the lumen. Bordering the core is the innermost layer of tissue called the tunica interna. The next layer is called the tunica media and primarily consists of muscle. This is smooth muscle, the last of our three types of muscle found in human bodies (the others as you recall are cardiac and skeletal). The outermost layer of an artery is called the tunica externa (or adventitia). It is primarily made of collagen and elastin, those same proteins found in skin.

These three layers must have several important characteristics in order to accommodate a heart that is pushing fluid through them. They have to be able to stretch when blood suddenly floods through a given area. They also must be able to snap back to their original positions when the need is over. This constriction and expansion is mediated by the muscles and exerted over the entire length of tubing. As we'll see, the inability to maintain this elasticity has important consequences for us as we get older.

Veins

As mentioned earlier, blood is not just a local food delivery service. It also serves as the garbage disposal system. This system has its own set of vascular highways designed to carry deoxygenated, junk-filled blood back to the lungs. These highways are composed of bluish colored vessels called venules. These venules eventually combine into larger freeways which are our veins. From there, the blood is taken to the right side of the pump-house. It is then sent to the lungs to get rid of waste (carbon dioxide) and pick up oxygen.

Veins have a slightly different structure from arteries. They are composed of the same three layers of tissue, but they have considerably less muscle and elastic tissue. They are still flexibile enough to stretch out when needed, and elastic enough to snap back to the original shape. But by the time the blood reaches a vein, it has lost a lot of its pressure. Consequently, veins don't need nearly as strong a wall as arteries.

Whether one examines tiny hummingbirds or humans, most creatures employ both beating hearts and vessels to carry the life-giving oxygen to their bodies. The largest creatures in the world, the whales, have systems very similar in design to terrestrial organisms (they are mammals, after all). Considering the incredible amount of tissue that must be fed, their ability to distribute oxygen is quite spectacular. And so is their equipment. The heart of a blue whale, the largest animal on earth, weighs more than half a ton.

Aging and the vasculature

With some of this anatomy and function in mind, we can explore the effects of The Clock of Ages on our cardiovascular system. The summation of the data points to two interesting ideas. First, there are senescence processes in the cardiovascular system that can be reversed. This can be done through external input, essentially life-style changes like diet and exercise. It also points to the fact that there are processes that cannot be overturned, regardless of what you do. These irreversible changes are discussed below.

How does the aging process affect the heart and blood vessels? Most of the data have been obtained by observing people of different ages in dynamic exercise. Two important findings demonstrate an overall change in function with age. As mentioned in the paragraph above, cardiac talents are greatly affected by what we eat and how we exercise. The studies summarized below have taken these ideas into consideration. They have explored the populations as they exist in a single time frame (so-called cross-sectional studies) as well as in the long term (so-called longitudinal studies).

The first change researchers on human aging have noticed has to do with heart rate. A reduction in the fastest heart rate attainable during maximal exercise occurs as we get older. The second change has to do with the maximum amount of oxygen our tissues can slurp up as a result of exercise. It decreases. This change has been shown to be linear across the ages. By the time we reach 65, we have lost 30–40% of our aerobic power compared to a young adult. These two pieces of data, when examined together, tell a single story. The heart continually loses its efficacy as a pumping machine in later adulthood (Figure 21).

The explanations for this change are multi-factorial. One major reason is a reduction in the capacity of the left ventricle to push large volumes of blood into the aorta and out to the rest of the body. This occurs because as we age, the wall of the left ventricle thickens (there is a general increase

HOW THE HUMAN HEART AGES

It all starts wth the left ventricle, located here.

As we age, the walls around the left ventricle thicken. The enlargement reduces the ability of the heart to pump large volumes of blood to the body. Three measures of cardiac fitness decline as a result:

Stroke volume diminishes. This number represents the amount of blood the heart can pump at each beat. As less blood goes through, this number goes down.

Cardiac output decreases. This number is calculated by multiplying the stroke volume by the pulse. The stroke volume is diminished in older hearts, reducing total cardiac output per unit time.

Oxygen consumption declines. This number is calculated by multiplying the oxygen extracted from blood by the cardiac output. As the output declines, so does the ability of tissues to be fed.

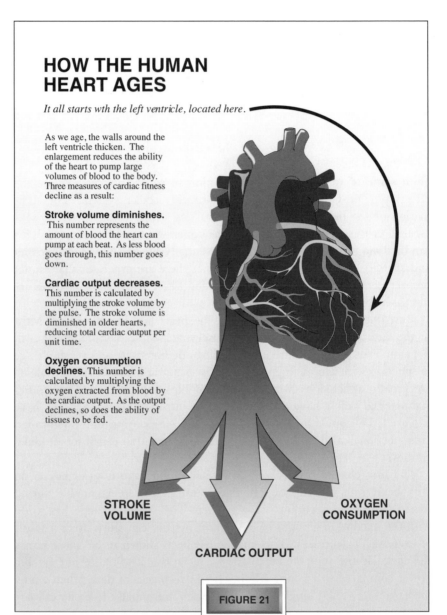

STROKE VOLUME

OXYGEN CONSUMPTION

CARDIAC OUTPUT

FIGURE 21

in overall ventricular mass). This thickening reduces the heart's potential ability to contract completely. More time is required to complete the filling of the atria, sending the blood to the lungs, returning the blood from the lungs and then pushing it out to the body. This reduces the stroke volume (as you recall, the amount of blood pumped out during a contraction) of an aging person. Since the stroke volume is used to measure total cardiac output (volume × pulse), this output also is lowered.

What does this slow-down do to the body? The exchange reduces the amount of oxygenated blood flowing through it at any one time. This loss means an overall decline in oxygen consumption by our tissues. During exercise, muscles are more likely to receive an inadequate supply of oxygen. This may be one of the reasons why as we age, we lose more and more of our aerobic power.

Time-out for an interesting exception

The idea that blood and nourishment had some kind of biological link occurred as early as the 17th century. The notion didn't have anything to do with oxygen, however. Consider this exchange in an English teaching hospital, being visited by the famed physician William Harvey: 'And how does the body replace the blood that it loses?' the teacher asked his 17th century medical class. 'Through the food we eat!' an eager young student bellowed. William Harvey rolled his eyes. The old idea was entrenched and still being taught: that blood was used up every time it was pumped to the body, and was replaced by the food eaten during the day.

Why did Harvey roll his eyes? The notion was quite obviously in error, giving the wrong impression about circulation to the entire class. Not to mention the entire century. Harvey had been able to prove it and had done so elegantly. This is how.

By performing an experiment on himself, William Harvey worked out the weight of blood that left his heart every hour. His calculations astonished him. They came to about three times the weight of his body. If the notion of blood replacement through food was correct, one would need to eat three times the weight of his body *every 60 minutes*. That's more than 450 pounds an hour for most males. A person could not perform such a feat in a day, or a week or a month. *Obviously* something was wrong. The conclusion he formed was the correct one: the same blood circulated round the lungs and the body time and time again. Recycling had to occur, and since the blood worked through blood vessels, the system had to be a closed loop.

Other reasons for aging

Even though Harvey discovered a closed blood system, the old idea that the health and amount of blood cells can be affected by what we eat has also proven true. Harvey might have been surprised to find that food consumption directly affects the manufacturing and the quality of the new blood we synthesize. He might also be surprised how deeply the aging process affects the distribution of this blood, newly manufactured or not. The reason is that the aging process affects not only the heart, but also that vast vascular highway Harvey brilliantly described so many years ago, the blood vessels (Figure 22).

As the years go by, there is an increased resistance to blood flow in our vasculature. The usual explanation has been that the walls of our arteries become more rigid. They are less able to expand when the heart sends a tidal wave of oxygenated blood down their corridors. Indeed, one can demonstrate a thickening in various tissues (notably changes in our old proteinaceous friend collagen as well as calcium deposits) in older arteries. This thickening could create a stiffer blood vessel and hamper its ability to contract or expand with age.

Even though an inhibition of expansion makes intuitive sense, the data concerning the outward thickening of blood vessels is somewhat controversial. The research is necessarily complicated by the disease of hypertension, whose incidence increases as we get older. This disease is characterized by a dangerous elevation in blood pressure and afflicts about 20% of American adults. There are several types, the vast majority occurring from circumstances that are still mysterious. Attempts to control for this disease in vascular studies have led to mixed findings regarding the overall elasticity of blood vessels in elderly people.

This thickening does not occur just on the outside of a blood vessel. It is a normal part of the aging process to accumulate molecules on a vessel's inner surface, thereby narrowing its internal diameter. As a result, it is thought that the heart has to work harder to get the blood to all the tissues that need it. There are many kinds of molecules responsible for this clogging. Among them are cholesterols, triglycerides and lipoproteins.

The data mentioned above were mostly measured when patients were

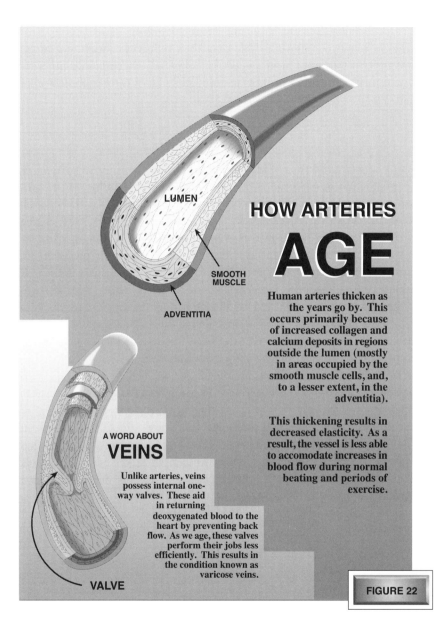

HOW ARTERIES

AGE

Human arteries thicken as the years go by. This occurs primarily because of increased collagen and calcium deposits in regions outside the lumen (mostly in areas occupied by the smooth muscle cells, and, to a lesser extent, in the adventitia).

This thickening results in decreased elasticity. As a result, the vessel is less able to accomodate increases in blood flow during normal beating and periods of exercise.

LUMEN

SMOOTH MUSCLE

ADVENTITIA

A WORD ABOUT

VEINS

Unlike arteries, veins possess internal one-way valves. These aid in returning deoxygenated blood to the heart by preventing back flow. As we age, these valves perform their jobs less efficiently. This results in the condition known as varicose veins.

VALVE

FIGURE 22

exercising heavily. During this time, a given person must pump great amounts of blood per minute. The data thus reflect the heart and blood vessels under a fair amount of stress. And they have revealed an inherent biological mechanism that cannot be completely altered by exercise (more on that later).

Conclusions

The fundamental observation is that the heart loses the ability to perform its primary function as the years go by – it cannot satisfy the oxygen needs of all the cells that require it. This incapacity has profound effects on every tissue of the body. For example, muscles need constant stimulation in order to keep the 'atrophy vultures' away from their proteins. To perform, these tissues need both neural signals and nourishing energy. If the blood cannot feed the relevant nerves properly, those neurons will die. That death signals the end of the ability of the muscle to respond to movement commands. So it starts to atrophy. If the muscle is insufficiently fed due to a poor vasculature, it will be just as inactive. And just as tempting a candidate for the atrophy scavenging system.

As we explore other systems in the body, we will discover just how important the affairs of the heart are. You might have guessed it really. Any deleterious changes to a process that provides system-wide vitality will profoundly reverberate throughout the body. These alterations in vascular processes have moved both peasants and kings; they have changed the complexion of society and altered the course of history. They have squeezed the tears from an English poet, who felt his heart break one lonely June morning in Florence.

8

The aging of the lungs

'I cannot continue to live in a world where there are beautiful blue-eyed, golden haired children. I cannot!'

These anguished words came from the lungs of one of the world's legendary dancers, Isadora Duncan. She was reacting to the painful loss of her own two children, Deirdre (aged five) and Patrick (aged three). They had drowned some years before, trapped inside a car that had plunged into the river. She uttered her despairing words often, whenever she saw living children that looked like her own little ones. Ironically, Duncan's own life would be cut short by a traffic accident too. And it would also involve a form of strangulation.

Isadora Duncan was born in San Fransisco in 1878. At an early age, she rediscovered the ancient Hellenistic ideas about freestyle movement and dance. Duncan was captivated immediately. She elevated these old principles, which were very much in contrast to the formal ballet regimen, to a creative artform for the 19th century. Duncan learned to coordinate voice with movement, spontaneity with grace and, like so many geniuses, innovation with ego. Early in her career, she remarked to a theatrical producer during an audition: 'I have discovered the art which had been lost for two thousand years. I bring you the idea that is going to revolutionize our entire epoch.'

She was not wrong, even in the opinion of her critics. Her dancing spread her fame around the world in giant leaps. And in several directions. Isadora Duncan began a series of torrid love affairs with artistic and wealthy men. These affairs would produce her children (different fathers) and make a reputation only a city like Paris could hold. To Paris she promptly moved.

It was in Paris that her children died. Taking a break between rehearsals, she lunched with Dierdre and Patrick and then hired a chauffeur to take them home. The car stalled on an embankment on the river

Seine. As the driver went out to recrank the car, he forgot that the transmission gear had been left engaged. The car rolled down the bank with the children still inside, slipped into the river and sank quickly. The helpless driver notified the authorities, but it was too late. When the car was raised, Dierdre and Patrick were dead.

Isadora Duncan was to know more tragedy in her life, including the horrifying suicide of the only man she would ever marry. Never very good with money, she found that that final fling with permanent relationships also meant the end of a steady income. She spent her last years as a near penniless dependent, living in Nice.

It was in this pauperish state that she would meet her demise. The dancer was walking along a main street when she spied a shiny red Bugatti sitting seductively in a show room. Always a lover of high performance vehicles, she could not resist going inside and at least asking for a test drive. Benoit Falchetto, the owner of the car dealership, recognized Duncan instantly. He volunteered to bring the car to her apartment next evening and allow a personally escorted trial drive.

It was cold that evening when the dealer arrived. Duncan, who was waiting outside, returned to her apartment to get a coat. She came out with an overlong, red shawl fringed with silk, and wrapped twice around her neck. After seating the dancer, Falchetto got into the driver's side. Unfortunately, the end of the dancer's long shawl dragged on the ground as the car accelerated. To the horrified screams of onlookers, the cloth got stuck in the spokes of the rear wheel, and immediately tightened its grip on Isadora Duncan's throat. She was jerked against the side of the car door; her throat was crushed in the process. Falchetto became hysterical, crying over and over that he had killed 'the Madonna.' The authorities were summoned, and while onlookers gawked, civic officials attempted to remove the body. The shawl was so tightly wound around the dancer's neck in this freak accident, however, that it had to be cut away. The police listed the official cause of death as strangulation.

Isadora Duncan was cremated in a few days and her ashes placed next to those of her children, Deirdre and Patrick.

The purpose of this chapter

Whether we are discussing the dancer's career, the deaths of her children or her own demise, much of Isadora Duncan's tragedies center on the functioning of the human lung. As mentioned in previous chapters, this organ seems especially vulnerable to the ravages of the times in which people lived. This is because our lungs provide the deepest conduit the outside environment has to the inside of our bodies. Any pathogen we gulp can follow the same conduit and deliver its destruction as deeply. And lungs are vulnerable to more than harmful micro-organisms. If a physical obstruction stops the lungs for as little as five minutes, irreversible damage to the brain occurs. Our lives are thus extremely vulnerable to any accident that even in the short term halts the function of the lung.

Unfortunately, the human lung is just as vulnerable to the aging processes. In this section, we are going to explore how the years change our lungs' capacity to interact with the planet's atmosphere. To discuss this process, we will first examine some anatomy and physiology. We will talk about blood and human attempts to measure how much oxygen red blood cells receive once they arrive in the lungs. Then we will discuss how The Clock of Ages affects this physiology, focusing on overall capacity as well as the performance of individual cells.

A brief tour of the human lung

The lungs, like every organ in the human body, possess amazing biological characteristics. They are divided into sections or lobes, three on the right side and two on the left. If you were to put your hands on them directly, these lobes would feel like moist sponges.

When you breathe, the air flows through the mouth and nose and down into the tube-like trachea. The trachea branches into two sections, one to each lung. These tubes in turn subdivide into millions of smaller tubes we call bronchioles. The whole thing looks as if your chest swallowed an inverted tree, no leaves attached, with myriads of tiny limbs branching into each lung. This complexity is important. The tree-like design creates an enormous surface area for breathing. If you stretched it all out, the surface area of the lungs would be roughly the size of a tennis court. And you need that much area to keep your oxygen-dependent tissues from starving to death (Figure 23).

Structure and blood

Each end of these tiny limbs possesses a structure that looks less like the tip of a branch, and more like, to mix metaphors, a blown-up bubble gum

LUNGS UNDER THE SURFACE

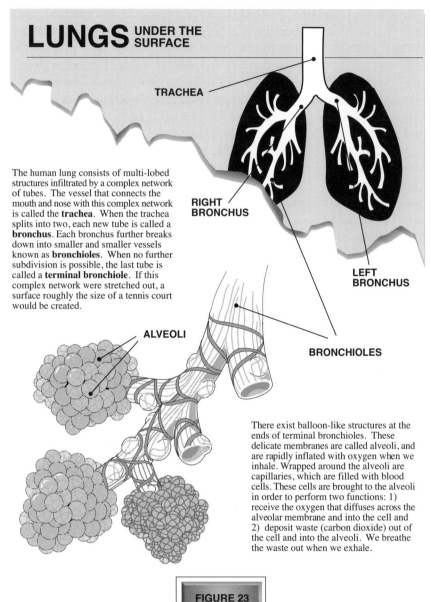

TRACHEA

RIGHT BRONCHUS

LEFT BRONCHUS

BRONCHIOLES

ALVEOLI

The human lung consists of multi-lobed structures infiltrated by a complex network of tubes. The vessel that connects the mouth and nose with this complex network is called the **trachea**. When the trachea splits into two, each new tube is called a **bronchus**. Each bronchus further breaks down into smaller and smaller vessels known as **bronchioles**. When no further subdivision is possible, the last tube is called a **terminal bronchiole**. If this complex network were stretched out, a surface roughly the size of a tennis court would be created.

There exist balloon-like structures at the ends of terminal bronchioles. These delicate membranes are called alveoli, and are rapidly inflated with oxygen when we inhale. Wrapped around the alveoli are capillaries, which are filled with blood cells. These cells are brought to the alveoli in order to perform two functions: 1) receive the oxygen that diffuses across the alveolar membrane and into the cell and 2) deposit waste (carbon dioxide) out of the cell and into the alveoli. We breathe the waste out when we exhale.

FIGURE 23

balloon. These objects are called alveoli and are some of the most delicate and important tissues you possess. They are very thin walled, only about 1/250 000th of an inch thick, and expand and contract with every breath you take.

Why are they so important? These balloons are surrounded by tiny capillaries, like a delicate hairnet might surround someone's head. The capillaries, as you know, contain blood cells, sent to the lungs from the heart. These cells have two very deep needs:

(1) They must receive oxygen; and
(2) They must get rid of waste.

Both of these needs are satisfied at the point where the walls of the alveoli meet the walls of the capillaries. The blood cell within the capillary gives up its waste (carbon dioxide gas) which passes into the alveoli. At the same time, the oxygen waiting in the alveoli (which got there the last time you breathed) is hurriedly shuttled in the opposite direction. The membranes act like revolving doors, with gases exchanged at a dizzying pace. The blood cell receives the oxygen and then returns to the heart. From there, now fully loaded with the gas, it travels to the rest of the body.

The lungs have a tricky balance to reach with the circulatory system. Blood cells aren't like a train that can wait around in the alveolar depot until oxygen decides to show up. The blood is continually being pumped through the body – and at a fairly fast rate. This means the oxygen has to be available at the time the blood cells show up or there will be no distribution. A delicate scheduling act has to take place so that the blood can dump off its toxic waste and take up the nourishing oxygen. As we'll see later, the aging process has a lot to say about this balance.

The mechanics of breathing

Our ability to breathe anything has to do with our ability to create a vacuum. As you know, air pressure is constantly beating down on our bodies, searching for vacuums to fill like a bloodhound looking for a fox. Most terrestrial organisms have found a way to exploit this atmospheric desire to fill space. We humans employ our diaphragm and certain chest muscles to contract, a phenomenon experienced every time we take a breath. This leads to an expansion, which in turn creates a vacuum in our lungs. The atmosphere around us encounters the vacuum and is immediately obliged to fill it. The gases come roaring down our trachea and into that gossamer framework of bronchioles and alveoli. The blood then picks up the new molecules and *voilà*. The exact opposite happens when we exhale. For a few minutes we create pressure that is greater than the atmosphere, and the gas in the lungs escapes.

Over the years, a number of mathematical measurements have been developed to measure the process just described. They can be grouped into two overall categories. The first category of numbers describes the factors that tell us the total capacity of a given lung. The other category describes the volume of gases that are moved in and out of our lungs during normal breathing. These numbers, like cardiovascular measurements, are obtained while the body is at maximal exercise. The results are then compared to data obtained while that same body is at rest. Such numbers give us a fairly detailed view of how our lungs are performing at any one time. And they give us an idea of how aging affects breathing.

If blood cells are less efficient at grabbing oxygen, it doesn't matter how good our twin bellows are working; the tissues won't get fed. If our blood cells are working fine but the tissue in the alveoli is shot, diffusion of gases may not occur. Even the walls of the capillaries that hold the blood cells are important (the gas has to travel through them as well; see Figure 23). So the ability to measure how well our lungs are working depends not only on overall measurements of capacity, but also on microscopic measurements.

A most unusual sled

This is not to say that our lungs can't take quite a bit of punishment. Sixteen times a minute they have to sample the atmosphere in which we reside, an atmosphere laden with other people's bacteria and viruses. But ignoring the pathogens for a moment, the lungs are resilient and strong. We can train our lungs to do enormous – even stupid – tasks. Consider, for example, the case of Jacques Mayol, a Frenchman swimming off the coast of Elba, Italy in December of 1983.

A special sled had been constructed in an attempt to establish the world's record for breath-held diving. Mayol was strapped to the sled and then was let go into the water. He descended a total of 344 feet in 104 seconds. He ascended from that depth in about 90 seconds. Total time of breath held? Three minutes, 23 seconds. His lungs, indeed his entire body, experienced changes in pressure that have never been felt by a human before or since. He set a world's record that day in December. And to date, no one has broken it.

It really makes you question Darwin when he comments that only the fittest creatures will survive.

Mayol's feat is only one example of the amazing ability of our lungs to withstand the stresses of life. Deep-sea pearl divers in the Pacific Ocean can hold their breaths for ten minutes and more. The record for a human escape from a damaged and sunken vessel with no equipment is 225 feet. You might recall the adventures of the American magician Harry Houdini. He invented a 'torture' device in which he was submerged upside down in a tank of water, bound and chained. It was his task to escape, which he did to the thrill of audiences everywhere.

The aging process

Whether you are a young Frenchman attempting a world's record or an American magician trying to make a buck, eventually the aging process catches up with your lungs. In this section, we will describe how some of the measurements mentioned earlier change as the years go by. We will start with a short chronology and then discuss what happens to various volumes and pressures achievable within the aging lung.

One of our first tasks after birth is to change from a comfortable fluid-filled environment to a cold, air-filled one. Making this transition requires the newborn to create a suction almost 50 times more powerful than the average adult breath. And then the infant makes up for lost time, averaging 40–60 breaths per minute. That rate changes as we get older, of course. At five years, we are breathing 24–26 times a minute. Post-puberty, the average pant is 20–22 breaths per minute, depending upon your company. It isn't until you get to the age of 25 or so that you achieve your adult rate, about 16 breaths per minute. And after that, the aging process begins to take hold.

What happens

There is an inescapable conclusion regarding lungs and age: move through the various stages of life, there is a reduction in the amount of oxygen that gets into the blood. We don't have the ability to grab the precious gas and drive it into our tissues like we could in a younger year. Even if you take into account diseases, lifestyle habits like cigarette smoking, increased exposure to environmental pollutants and other variables, the edict of The Clock of Ages is straightforward. There are many reasons for this decrease, and we will talk about a few of them below (Figure 24).

The first reason for this decrease can be described as an atmospheric parking problem. As we get older, it is harder for the lungs to distribute the air molecules equally throughout its labyrinthine territory. Not that it was always equitable. Because of gravity, blood was always more freely distributed in the lower part of the lung. Even in our youth, the more air that got down there, the more oxygen the body would receive.

The aging process dictates that less air goes to those lower places as the years go by. As a result, there is less oxygen to distribute to the blood which is already resident. What does that mean? *It means that some blood goes away without ever getting any oxygen at all.* Poorly oxygenated blood means more poorly fed tissue. This atmospheric distribution problem has system-wide negative effects.

You might be asking, as are many researchers, the following question: why do the lower parts of the aging lungs refuse to give oxygen a parking permit? The answer has to do with atmospheric squatters. And to a phenomenon known as elastic recoil, which is described next.

Elastic recoil

Elastic recoil is the tendency of a lung to resist expanding as air enters it. That's a good thing. If the lungs didn't resist, they'd keep right on getting bigger and bigger with every breath until they burst. This resistance produces a positive pressure, of course, and the resulting tug-of-war keeps our airways open. This same recoil helps to keep the airways unobstructed when we exhale, as well. That way the lungs only 'collapse' and empty their load when the respiratory muscles tell them to do so. There is a balance achieved between the pressure our lungs experience (because of the air inside them) and the muscles that supervise inhalation and exhalation.

What does this have to do with atmospheric squatters? If the airways closed before you finished exhaling, air would be trapped inside your lungs. When the next breath occurred, the lungs would still be filled with the previous breath's air. The new stuff would have nowhere to go. And the oxygen level in your blood would go down. You would begin 'strangling,' not because there was a scarf around your neck, but because the blood would not get any new supplies of fresh atmosphere.

This is exactly what happens when we age. Elastic recoil is lost. The top part of an older lung can still fill up and exhale. But the critical bottom part – where much of the blood goes – does not fill up readily, if at all. The elastic recoil is reduced below the crucial point needed to keep the airways open when you exhale. As a result, old air gets trapped. And the prior occupancy means that blood gets less oxygen. Less oxygen to the tissues means weakening occurs, especially when we consider the great demands our muscles place on our vasculature. That's why, as we get older, it is increasingly difficult to do heavy work.

Why do we have to lose elastic recoil?

The fault is in the alveoli, those bubble gum balloons that interface with our capillaries. For some reason, The Clock of Ages dictates that we lose a number of these elastic end pieces as we get older. It also decrees that the ones that are left change their composition, primarily in their internal amounts of good old elastin and collagen. As a result they become thicker and less flexible. These changes all contribute to a loss of elastic recoil. And why *those* changes occur in a chronologically predictable fashion is unknown.

There are other mechanical and biological reasons why we extract less and less oxygen from the air we breathe. There are problems with muscles that run our bellows, interactions with nerves that command these muscles, changes in the vessels that give them blood, and so on. The effect of aging on our lungs is best understood as multi-factorial. And it is quite dramatic. By the age of 70, this oxygen-to-tissue transfer rate has been halved.

Right now, we can only marvel at our breathing apparatus. And be profoundly shocked at the ease of its loss, whether we are considering the 19th century drownings of little children, the 20th century freak accidents of their parents, or the aging inside our own chests.

THE AGING LUNG

It's the lungs' fault that our blood gets less oxygen as we get older. Here's what happens.

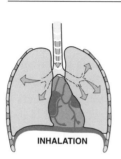 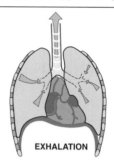

INHALATION

The muscles in our chest expand the volume of the lungs, creating a partial vacuum. Air flows in as a result of this expansion.

EXHALATION

The chest muscles then reduce the volume of the lungs. The change in pressure squeezes air out of our bodies.

WE START WITH THE FLUSHING OF AIR

Breathing occurs because the muscles in our lungs create a vacuum (see panels on the left). During exhalation, it is important to evacuate as much air from the lungs as possible. This results in a more efficient flushing of waste products. It also creates more room for fresh air to be delivered to the blood supply.

WE END WITH AN EXCHANGE OF MOLECULES

Red blood cells arrive at the alveoli via capillaries. These cells are laden with carbon dioxide waste.

1) The carbon dioxide diffuses from the red blood cell. It crosses the membrane of the capillary, and enters the alveoli. From there it is exhaled from the body.

2) The oxygen diffuses from the alveoli across the capillary membrane. It enters the red blood cell, which will transport it throughout the body.

1 CARBON DIOXIDE INTO THE ALVEOLI

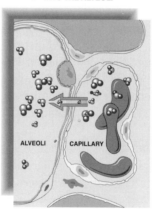

ALVEOLI   CAPILLARY

2 OXYGEN INTO THE RED BLOOD CELL

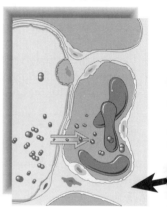

FIGURE 24

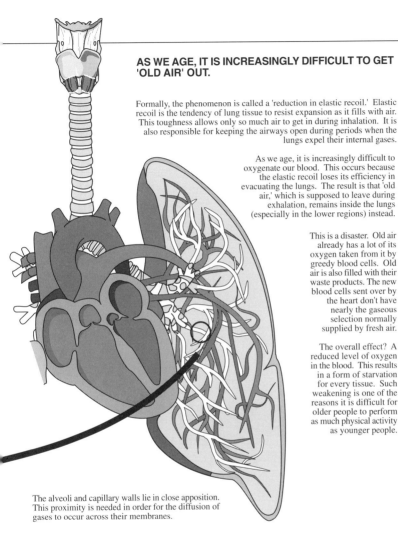

AS WE AGE, IT IS INCREASINGLY DIFFICULT TO GET 'OLD AIR' OUT.

Formally, the phenomenon is called a 'reduction in elastic recoil.' Elastic recoil is the tendency of lung tissue to resist expansion as it fills with air. This toughness allows only so much air to get in during inhalation. It is also responsible for keeping the airways open during periods when the lungs expel their internal gases.

As we age, it is increasingly difficult to oxygenate our blood. This occurs because the elastic recoil loses its efficiency in evacuating the lungs. The result is that 'old air,' which is supposed to leave during exhalation, remains inside the lungs (especially in the lower regions) instead.

This is a disaster. Old air already has a lot of its oxygen taken from it by greedy blood cells. Old air is also filled with their waste products. The new blood cells sent over by the heart don't have nearly the gaseous selection normally supplied by fresh air.

The overall effect? A reduced level of oxygen in the blood. This results in a form of starvation for every tissue. Such weakening is one of the reasons it is difficult for older people to perform as much physical activity as younger people.

The alveoli and capillary walls lie in close apposition. This proximity is needed in order for the diffusion of gases to occur across their membranes.

9

What happens to the digestion

Even if it's not true, you have to admit, it makes a great story. Ludwig van Beethoven was in bed, suffering from the diseases that would take his life. His abdomen was swollen with fluids, his liver in a severe state of dysfunction, his lungs losing a battle with pneumonia. And of course, his ears, filled with the unwelcome cement of otosclerosis, colored all this suffering in the opacity of silence. Beethoven was a medical wreck.

These multiple biological assaults were a familiar theme in Beethoven's life, even in his younger years. He probably suffered from lupus erythematosus, an autoimmune disease in which soldiers of the immune system suddenly attack healthy body cells. He also suffered from continual diarrhea and a variety of intestinal disorders, so severe that its pain often prostrated him. This combination of diseases culminated in periodic, debilitating rounds of depression. These the composer called 'as great an evil' as his hearing loss. All this pathology made him a grumpy, cantankerous man in his later years.

In 1815, the now deaf composer was awarded the guardianship of his nephew, a child of Beethoven's recently deceased brother Caspar. Though there was great love between them, they fought constantly. Just before taking his university entrance examinations in 1826, the boy shot himself. Horrified, Beethoven bundled up the wounded nephew, went out into a terrific winter storm and dropped him off at the home of a relative. The grieving composer returned to his own house in an open cart, inadequately clothed. Beethoven was ill in a week, and never fully recovered.

The morning of March 26 would be Beethoven's last. With pneumonia, cirrhosis of the liver and intestinal disorders all recapitulating their violent strains, the composer awoke in a stupor. Appropriately, another thunderstorm raged over the skies of Vienna that day, shattering windows and drenching the town. Beethoven asked for a glass of wine, and when it was slow in coming, uttered his last words, 'Pity, pity – too late.' He slumped into unconsciousness.

All of this is fact. The legend starts at the point just after his final words. There apparently was a large succession of thunderclaps. One occurred right over the house in which Beethoven lay. It was of such violence that it shook the frames and floor of the building, vibrations Beethoven could perceive. He awoke suddenly from his coma, wild-eyed and furious. He sat bolt upright, lifted his right hand to the heavens in a clenched fist, shook it for all its worth and, exhausted, fell back into bed. And died.

The next section

Like a fog, legends have a way of obscuring the truth about great people. Or at least de-emphasizing important aspects of their life. Beethoven's sufferings are like that. Though the composer's hearing loss is the disease one mostly associates with his dark manner, his digestive problems contributed greatly to his moods and dispositions. Many of them centered around his cirrhosis of the liver. Though he was not the town drunk, he was known to binge. His father, a Flemish man, was apparently a severe alcoholic.

I would like to use the intestinal problems associated with Beethoven's later years as a starting point for the subject of this chapter. We are going to be talking about aging digestive systems, not just in famous composers, but in all of us. To do this, I will guide us on a tour of the digestive tract and how its various components change as we get older. We will trace the ingestion of a particular food through the primary organs of digestion, commenting on particular senescence processes as we go. We will talk about excretory systems, both bowel and bladder, and discover that some tissues weather the aging process better than is commonly believed.

As ever, when we discuss digestive degeneration, we must be aware of the effects of dietary habits, lifestyles and prematurely acquired disease. Enough data exist for us to see through this thicket of variables, however, and it is that work which will be discussed.

The oral phase of digestion

Our discussion of gastrointestinal aging necessarily includes a description of some normal anatomy and physiology. We will illustrate these processes by following the digestion of what may have been one of Beethoven's favorite foods, a calorie-laden Viennese chocolate torte (Figure 25).

When the conglomeration of carbohydrates and fats that exist in a torte enter your mouth, the tearing and grinding actions of your teeth smash it into an unsavory goo. Its presence also induces the secretion of saliva, the first encounter with chemical digestion. Saliva not only helps digest various foods, it also possesses molecules that protect our mouth and gums from bacterial infection. By dissolving, lubricating and washing away food particles, saliva allows the various flavors maximum access to our taste buds. It is quite a fluid.

After the mouth has finished grinding the torte into a partially digested slurry, the food is swallowed. As you know, the tube that receives it is called the esophagus. You may not know that a reflexive series of muscles are called into action to aid the food as it goes down your throat. Their collective action is called peristalsis. This is a wave that progressively travels from the top of the esophagus to the top of the stomach. It is powerful enough to transport food even if you are lying down. To get into the stomach, the remains of the torte have to pass through a narrowing area called the esophageal sphincter. This part of the esophagus acts like a gatekeeper, allowing the slurry into the stomach and then closing up quickly. The powerful muscle prevents the stomach's caustic acids from escaping back into the throat. When the sphincter doesn't do its job and some of those acids splash into the throat, we experience heartburn.

How it ages

As we get older, various components of the oral digestion system begin to change (Figure 26). The parotid glands, responsible for secreting saliva into the mouth, make less fluid. As a result, people in later adulthood tend to experience the symptom of dry mouth. This reduction in salivary secretion means a decreased protection of the teeth, tongue and gums from bacterial infection. It also results in greater discomfort when talking and a greater difficulty in chewing foods – especially if the foods have

HOW TO DIGEST A CHOCOLATE TORTE

Food passes from lips to loo in about 8 hours. Here's what happens if you eat a dessert at lunch time.

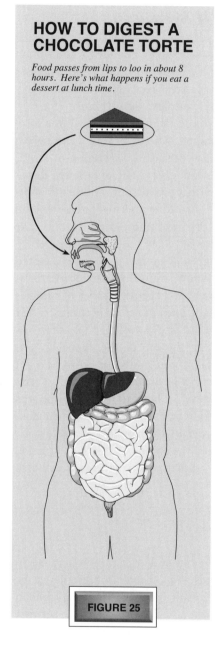

MOUTH
The torte is chewed by the teeth and partially digested with saliva. The tongue throws the remnants to muscles in the esophagus, which facilitate swallowing.

STOMACH
Acids in the stomach break the swallowed chunks into a soupy mixture called chyme. Stomach contractions, occurring every 20 seconds, continually churn the mixture.

SMALL INTESTINE
The soupy chyme travels to the small intestine. More contractions occur which further break down the biochemicals in the torte. Subsequent nutrients are absorbed through tiny blood vessels in the intestinal tissue.

PANCREAS
This organ secretes substances into the chyme via ducts that connect to the small intestine. These fluids help to neutralize residual stomach acid and also aid in the digestion of fats, sugars and proteins.

LIVER
Liver makes fat-emulsifying bile, which is also dumped into the chyme. The bile is stored in the gall bladder until needed. The liver also detoxifies the food-laden blood.

LARGE INTESTINE
Undigested food enters the large intestine, where excess water is removed and returned to the body. The residue is pushed into the rectum for elimination as feces.

KIDNEY
The blood is recycled through the kidneys, which act as purifying filters. Collected waste is turned into urine and stored in the bladder until it can be eliminated.

ELIMINATION
Approximately 150 grams of feces are eliminated per day. They consist of about 100 grams of water, and 50 grams of bacteria and any undigested parts of the torte eaten the night before.

FIGURE 25

very little moisture. Because saliva also spreads flavors around tongue surfaces, a loss of fluid also means a reduction in taste sensation.

The data on the effects of the years on our mouths is fairly solid. In contrast, the data on the aging of our swallowing mechanisms has been filled with controversy. Some studies suggest that the esophageal muscles, those tissues responsible for peristalsis, get weaker with age. This may in part be due to an overall weakening of muscle tissue, as described previously. There is some evidence that the intensity of the wave created during swallowing is diminished. Other studies have suggested just the opposite, that we retain the ability to swallow almost anything with the same vigor in our older years as in our younger ones. As of this writing, the question is mostly unresolved. In some people there are dramatic changes; in others, none at all.

What happens next?

After the torte enters the stomach it is no longer recognizable. It has passed through your mouth, and has become a chunky semi-fluid in the process, depending on how well you chewed. It was tossed down your throat and was allowed entrance to the stomach via that sphincter. It is now ready to be pulverized by the amazing substances your stomach secretes.

The way the stomach digests food can be thought of like a washing machine in which strong acids and other fluids are continually being dumped. The food is agitated even as it is being broken down. Gastric juice, the formal term for the fluid, contains not only acid, but also proteins, mucus and hormones. They are all designed to blast the chunky slurry it received into a smooth liquid that your intestines can utilize. By the time they're done, the stomach has transformed the torte into slurry, which formally goes by the name of chyme.

Does it age?

The answer is yes. The effects do not appear to cripple function, however. There are changes in the functionality of the muscles that govern the washing machine aspect of digestion. It is thus more difficult to agitate the food that enters the stomach. There is a decline in the amounts and

How the digestive system ages

The tissues that feed us deteriorate unevenly. Some organs exhibit no real loss of function in our later years, whereas others show dramatic changes. Summarized below are some of the alterations our major organs of digestion undergo. On the next page, the aging of the small intestine is discussed.

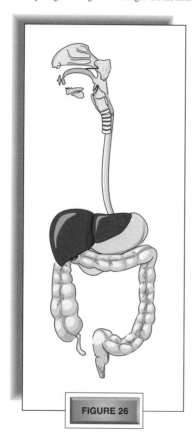

FIGURE 26

SALIVA
The parotid glands secrete less saliva as we age. This results in a drier mouth and greater discomfort in talking and chewing.

SWALLOWING
The muscles responsible for swallowing weaken. This deterioration is not significant, however, and no loss of function occurs with age.

STOMACH
Muscles weaken and gastric juice contents change with passing years. Except for difficulty in digesting meats, no real change in function occurs.

LIVER
There are alterations in both cellular numbers and overall architecture of various tissues within this organ. These do not result in functional changes.

PANCREAS
There is a reduction in the fat-eating molecules secreted into the intestine. The liver can adequately compensate for this loss, however.

LARGE INTESTINE
A reduction in lubricating mucus and a weakening of the muscles which push waste into the rectum occurs. No loss of function occurs.

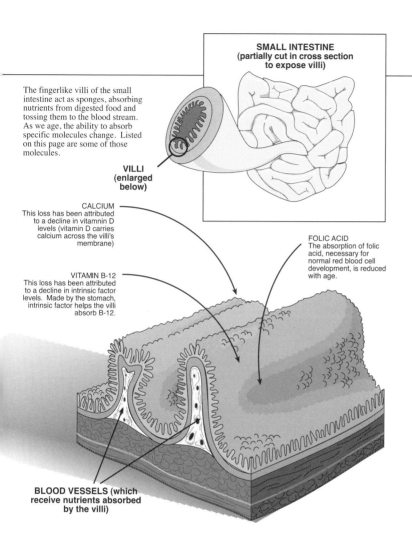

**SMALL INTESTINE
(partially cut in cross section
to expose villi)**

The fingerlike villi of the small intestine act as sponges, absorbing nutrients from digested food and tossing them to the blood stream. As we age, the ability to absorb specific molecules change. Listed on this page are some of those molecules.

**VILLI
(enlarged
below)**

CALCIUM
This loss has been attributed to a decline in vitamnin D levels (vitamin D carries calcium across the villi's membrane)

FOLIC ACID
The absorption of folic acid, necessary for normal red blood cell development, is reduced with age.

VITAMIN B-12
This loss has been attributed to a decline in intrinsic factor levels. Made by the stomach, intrinsic factor helps the villi absorb B-12.

BLOOD VESSELS (which receive nutrients absorbed by the villi)

components of the gastric juice also. By the time we reach the age of 60, we've lost about 25% of the volume we normally secrete for a given volume of food. There is a 60% decline in the activity of a gastric molecule called peptin, used primarily in the breakdown of proteins. As a result, it is more difficult to digest heavy meats in later adulthood. But even considering these dramatic changes, other foods are broken down without noticeable discomfort. The Clock of Ages does affect certain aspects of our stomach's biology, but it does not appear to interfere with much of its overall function.

On to the small intestine

Now that the torte has been changed into a semi-smooth soup, it is ready to enter the 22-foot-long small intestine. Here is where your food actually becomes broken down into the tiny molecules your body utilizes to stay healthy. It is where true 'feeding' occurs, given friendly assistance by the liver and pancreas.

The internal tube of the small intestine looks like a long underwater cave filled with coral. There are many fingers that project into the hollow space of the tube. These projections greatly increase the internal surface area of the small intestine. Called villi, they act more like sponges than coral. They absorb any nutrients they find in the chyme. This absorption can take some time, however, and the intestine has found an ingenious way to allow maximum exposure of the villi to its food source.

It dances. Or, rather, moves in a rhythmic pattern. Once it receives the chyme, circular muscles within the tube begin to contract and expand at various intervals. This has the effect of sloshing the fluid backward and forward along the length of the tube. Termed segmentation, this process ensures maximum exposure of those little finger villi to the nutri-ent-laden soup. It also allows time for secretions from other organs (notably the pancreas and liver) to enter the intestine and perform useful work on the chyme. These aren't random motions, there is basic forward movement of the soup through all 22 feet of small intestine. But this action allows the villi to dine on what was left of the torte at leisure, extracting every available nutrient.

How it ages

Unlike the stomach, there are specific changes in the ability of the small intestine to perform its function as we age. Some of these changes have to do with the muscles and the large tissues of the tube itself. Others are concerned with the ability of the villi to pick out certain molecules within the chyme. Each are discussed below.

As we get older, the physical weight of the small intestine decreases. There is also a reduction in the absorbing surface area inside the tube, which may be related to the decreased weight. Why does this occur? Those finger-like villi get broader and shorter as we age. Eventually, they will stop looking like individual coral projections and start looking like parallel mountain ridges.

Do those changes *really* do anything to our digestive capacity? Probably. But the overall effect may be reduced, because of the average life span of cells within a single villus. They turn over every 3–5 days, even in our older age. That's a high rate. The aging process may pose very little threat to cells which never really 'age.' Moreover, we have so many villi within the small intestine that even when flattened, they still may be able to extract most of the nutrition we need.

The real effects of aging are seen not by considering functions of the overall tube, but by examining the ability of villi to absorb specific molecules. For example, after the age of 70, our ability to absorb calcium is severly reduced. This may be due to a loss of vitamin D, a chemical needed to grab calcium from the chyme and stick it into the blood stream. There are researchers who believe this loss is the reason our bones become demineralized as we get older.

The villi seem to lose other functions with age. Vitamin B$_{12}$, for example, is not as easily absorbed from the chyme. This may be due to a loss of something called intrinsic factor, normally secreted by the stomach. Vitamin B$_{12}$ is important for energy production, red blood cell formation and the creation of certain neurotransmitters. Its loss may contribute to certain deficiencies observed in old age.

There are many other changes associated with aging and our ability to digest food, and the list seems to be growing yearly. It must be observed, as ever, that lifestyle habits, diet, even the ingestion of medications can obscure the natural process of digestive degeneration. It does seem certain that The Clock of Ages exerts its force over the small intestine. This results in changes observed not only at the larger tissue level, but also at the level of the molecule.

The organs that help the small intestine

The digestive process within the small intestine could not be completed without some assistance from other organs. The reason is that the small intestine is incapable of breaking down everything in the soup that it receives. The liver and the pancreas help the small intestine bust up the food as it chugs along the tube. These organs are also subject to the aging process.

The fats in the torte are broken down by substances originating in the liver. They perform this delightful task by secreting yellowish bile directly into the small intestine. Bile acts a little bit like soap, emulsifying fats into a form that the villi can absorb. The yellow stuff also activates fat-munching molecules produced by the small intestine itself. Together they create bite-sized molecules the villi absorb with ease. When you are not eating substances like tortes, bile is stored in the gall bladder.

The pancreas is another organ that helps the small intestine to absorb the molecules in the chyme. Like the liver, it has direct access, dumping two kinds of torte-munching substances into the tube. One kind contains a fancy concoction whose active ingredient is baking soda. This substance helps to neutralize all the acids the chyme received from the stomach and dutifully carried into the small intestine. The other substance is a mixed bag of food-eating molecules. These include chemicals that will break down proteins, sugars and, if the bile needs any assistance, also fats. The combined action of these substances from the pancreas, the bile from the liver and the movement of the small intestines, assures us that we will wring out, like a dishrag, every important molecule the chyme possesses.

And, considering the calories in the chocolate torte, even a few we would rather they left alone.

Aging in the liver and pancreas

Can the aging process be observed in the organs that help the small intestine digest food? And if it can, does it really matter? The answer to these questions is yes and probably no. The reason has to do with some of the back-up properties in our liver and pancreas.

One can observe age-related cellular changes in size and structure of various tissues in the liver. But they don't seem to affect its function. We still secrete adequate amounts of bile as we age, and there are no age-related changes in various blood-borne chemicals the liver normally makes. There was a time researchers thought that total liver weight declined with advancing years. That has been shown to be false.

One might expect such steadiness from an organ that has properties very different from many in the body. You can remove up to 80% of the organ and not seriously affect your health (try doing that with your heart). In addition, the liver has this odd ability to regenerate itself. This replicative ability probably compensates for any loss The Clock of Ages might throw its way.

The aging process does not seem adversely to affect the pancreas either. As in the liver, one can notice some age-related differences in structure. But these differences don't appear to alter its critical digestive functions, except in one case. It has been shown that the amount of some fat-eating molecules the pancreas secretes into the small intestine decrease with age. But no matter. The liver can more than compensate for this loss.

So far, it seems as if the digestive system is holding up well in our maturity. We can chew, swallow and break down large chunks of food in the stomach. We are able to squeeze nutrients out of this pre-digested material, even if the selectivity changes with age. But is the digestive system equally hardy when it comes to getting rid of the waste? After the small intestine and all the various organs have exerted their force on the torte, it is time to get rid of excess solid and liquid matter. These humble tasks are left to the large intestine and kidneys, organs we consider next.

The colon

The remnants of the torte after the small intestine is done digesting will never be used by the body. The last big task is to absorb the water in the chyme. This will serve to solidify and compact the waste material, as well as recycle precious liquid. These actions are supervised by the large intestine, which receives the chyme directly from the small intestine.

The large intestine performs this consolidation in a manner similar to that in the small intestine. It slowly agitates its contents back and forth, creating the motion through a process resembling segmentation. There are tissues that allow water to pass out of the intestine. The greater the time the residue can slosh through the large intestine, the more water is absorbed. The motion is in a single direction, however, like a conveyor belt for fecal material. The increasingly dry material will eventually reach the rectum, where it is stored until it can be expelled. If this motion is halted, the person is said to be constipated.

But does it age?

When I was in junior high, I noticed something very peculiar about certain TV ads. Whenever a laxative commercial was shown, the host and subjects of the spot were always elderly. Invariably there would be this glum-looking person – usually a woman – who was not feeling 'regular.' I soon found out this term was a euphemism for having immobile fecal material in the colon. I observed this tendency to use older people in these spots even into my undergraduate days. I thus grew up with the impression that the older you got, the more constipated you became. This was thought to occur because of a change in the ability of the colon to pass along waste products.

But does this prejudice have any basis in reality? The surprising answer is no. *There do not appear to be major functional losses in this organ as we get older.* The commercials were and are wrong. The colon, instead of degenerating into a laxative-begging bag of cells, is rather just like most of the rest of the digestive system. If kept in fair maintenance, the tissues work for a very long time.

There are some structural changes that occur in the large intestine, primarily at the cellular level. There is a reduced secretion of mucus, a substance that helps in lubrication. There is some degeneration of the muscles responsible for movement of fecal material into the rectum. None of these changes can be associated with alterations in fecal habits. When constipation occurs in an elderly person, it is generally due to some other reason than The Clock of Ages.

The kidneys

The next organ we consider in the digestive process is also the last stop on our torte tour. The blood is filled with new food-related molecules, freshly extracted from the torte. The blood now needs to be filtered and the toxic waste removed. As you know, this job belongs to the kidneys, those bean-shaped organs in the back of the abdomen. We will discuss a little bit of their anatomy and then talk about their aging process. They are the one organ in the digestive system The Clock of Ages does not leave alone.

The operative toxic waste inspector in the kidneys is called a nephron, which is divided into several parts. The blood first enters a nephron through a large blood vessel. It immediately encounters a round structure called Bowman's capsule, and, just as immediately, the blood vessel breaks into a tiny spider's web of capillaries. It looks like a tree trunk that has branched into a million little roots.

This root system is where the filtering action occurs. The capillaries release their wastes into a series of collecting tubes inside the nephron. These tubes send the fluid, which is now called urine, through a tubular tangle of ducts. Eventually the urine is brought to the bladder through tubes called ureters, and here it waits to be eliminated. Whereas the digestive process can take many hours to complete, the entire food-laden blood supply is cleansed about every 30 minutes.

Do kidneys age?

Unlike other digestion-related tissues in the body, the kidneys really do exhibit signs of deterioration in later adulthood (Figure 27). The first hints of changes show up in the overall mass, starting at about age 30. The volume and weight of the kidneys shrink as we get older. The area that seems particularly affected is the one housing those round, capillary-containing Bowman's capsules. The kidney's overall appearance also becomes smoother, reducing the aggregate surface area of the filtering apparatus.

There are more shape changes when one explores the internal tissues of the kidney. The 'tree trunk' and 'root system' vessels that bring the blood cells to the collecting tubes of the filtering system begin to stiffen and narrow. Those collecting tubes that receive the wastes from the blood begin to change their appearance, too. They get thicker. They also shorten their length. They even develop unwanted forks and shunts, which create little pools that collect debris. These changes occur because there is a net

Changes in elimination

Our ability to hold urine and cleanse blood changes as we get older. Here's what happens to our urinary tract and kidneys.

The structure of the bladder and kidney changes as the years go by. These changes can have a dramatic effect on lifestyle and health. Discussed below left are events that occur in the bladder. Below right and on the next page, the kidney is discussed.

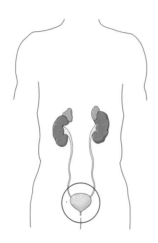

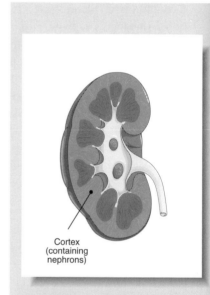

Cortex
(containing
nephrons)

THE BLADDER
As we age, the connective tissue and muscles permitting expansion and contraction of the bladder weaken. This results in an overall decrease in our ability to store urine. It also reduces the efficiency of emptying the bladder when urinating.

THE KIDNEY
The weight and volume of the kidneys shrink as we age. Their appearance becomes smoother, reducing the aggregate surface area for filtering. These changes reflect alterations occurring within the nephron, shown next page. This structure is composed of blood vessels. The blood is filtered there. Nephrons are located in the cortex of the human kidney as shown above.

FIGURE 27

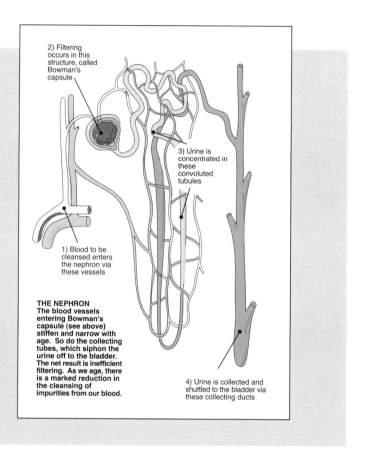

2) Filtering occurs in this structure, called Bowman's capsule

3) Urine is concentrated in these convoluted tubules

1) Blood to be cleansed enters the nephron via these vessels

THE NEPHRON
The blood vessels
entering Bowman's
capsule (see above)
stiffen and narrow with
age. So do the collecting
tubes, which siphon the
urine off to the bladder.
The net result is inefficient
filtering. As we age, there
is a marked reduction in
the cleansing of
impurities from our blood.

4) Urine is collected and shuttled to the bladder via these collecting ducts

loss of certain cells within these structures. Why those cells die remains a mystery.

What effect do these and other changes have on overall kidney function? Because of vessel impairment, blood flow through the kidneys is increasingly hampered as we age. There is consequently less blood available for cleansing at any one time. That means the waste products are not as effectively removed from our tissues as we get older, and their presence portends system-wide damage. The thickening of those collecting tubes means a decreased ability to snatch up toxic substances as the blood goes by. Less waste in the bladder means more waste in the blood. The problems started by the aging blood vessels are aggravated by the aging tubes. By the time we reach age 90, we can lose almost half of the filtering capabilities we had at age 20.

The effects of dysfunction in this system can be seen in the death of the great inventor, Thomas Edison. He was famous for his partial deafness, but also suffered from diabetes, gastric ulcers and a condition known as Bright's disease (also called glomerulonephritis). The disease causes a large build-up of fluids in the body and severe back pain. The effects of his Bright's disease on Edison's health were dramatically observed in October, 1929.

Edison had been invited to be Henry Ford's honorary guest in Dearborn, Michigan. There was to be a celebration of the 50th anniversary of the electric light. Herbert Hoover, the President of the United States, was in attendance and he had just finished his opening remarks. An 82-year-old Edison was motioned to take the podium and as he got up, the inventor felt something terribly wrong. He collapsed in a heap. He was taken back to his home in New Jersey, where he steadily grew weaker.

He never fully recovered. Uremic poisoning was the cause of his failing health, due to an inability to filter the wastes his blood received. He died in 1931.

Let's not leave the bladder out

This filtering system is thus extremely important to the maintenance of good health, regardless of who you are. But filtering is hardly the entire story. The muscular bag that holds the urine sent by the kidneys is also very important to the elimination of toxic wastes – and just as subject to the aging process. To understand what happens, we have to explore the overall mission of this interesting organ. The bladder is at maximum operating capacity when it can perform two functions:

(1) Expand to hold urine between visits to the bathroom without discomfort; and
(2) Empty out completely when voiding.

Unfortunately as we get older, both of these characteristics begin to change. And not for the better. The problem is that the connective tissue containing that all too familiar group of proteins, collagen and elastin, begins to degenerate. As a result, its ability to expand and contract begins to weaken. What this means is that the total amount of urine which can be stored decreases as we get older. It also means that it does not empty out as well.

Another age-related change may have more to do with nerves than with muscles. There are reports that the perception of the need to void the bladder changes as we get older. In young adults, the reflex that tells us it is time to urinate occurs when the bladder is about half full. In people over the age of 65, that message may not be perceived until the bladder is nearly full. There may not even be any sensation to go at all. As we age, we have less advanced warning that we have to use the bathroom. And considering the volume, we have more need to use it.

Taken together

As we observe the passage of years over our digestive systems, the perception that the aging process is capricious and arbitrary is almost irresistible. Why do the cells in the stomach, which are subject to extremely hazardous conditions, work so efficiently well into our later years? Why also the small intestine? The colon? What is so different in the cells mediating kidney function that the organ is rendered only half as effective? Or the bladder?

When we compare how the digestive system ages to how other systems age, we are struck with the same research need. The obvious answers to the aging process lie in the cells of the organs we study. We can hypothesize and ultimately discover the secrets of aging by comparing cells which do not appear to break down with those that do so easily.

This was the struggle so mightily performed in Beethoven's aging body. While his mind seethed and raged with the youthful and titanic struggles of musical genius, his supporting digestive system bore the brunt of his aging years. He probably still had many symphonies left in the nerves of his mind. He just, after almost six decades of life, ran out of organs that could feed them.

10

How the senses age

It was not the most welcoming way to be born into the planet.

The doctors thought that Samuel Johnson, the great author and conversationalist, was still-born. Fortunately for Western literature, he was not abandoned. The future writer was shaken, slapped, cajoled, yelled at and paddled until he convincingly shrieked his vitality into existence. The events of his birth were simply a foreshadowing of a troubled medical future, however. Samuel Johnnson would not die until the age of 75, and not until he had experienced most of the known ailments and diseases of his day.

As a baby, Samuel Johnson contracted tuberculosis and developed a glandular infection of the neck. In his later years he suffered asthma, dropsy, emphysema, gallstones, gout, hydrocele (swelling of the scrotum), manic depression, a diseased bladder which left him incontinent and chronic bronchitis. 'My health seldom afforded me a single day of ease,' Johnson wrote about his troubles.

His senses were not left untouched, either. His bouts with disease left him blind in one eye and partially deaf. There was a point when Johnson suffered aphasia, a temporary loss of speech, probably due to a cerebral hemorrhage. The only sense that seemed to amplify with the passing years was a constant feeling of pain.

Despite these sensory losses, Johnson was able to live through the better part of seven decades, an astonishing achievement for anyone living in the 18th century. The end came in December of 1784, the fluid in his lungs reaching a level that drowned him. His biographer Boswell wrote that Johnson entertained and conversed with company to the bitter end. And the end soon came. 'His difficulty of breathing increased till about 7:00 in the morning,' Boswell wrote, 'When Mr. Barber and Mrs. Desmoulins, who were sitting in the room, observing that the noise he made breathing had ceased, went to the bed, and found he was dead.'

The power of living

Samuel Johnson's life is a great tribute to the will to survive hostile conditions, and the ability to be productive in the face of increasing sensory deprivation. I would like to use his remarkable ability to adapt to his conditions as a point for a discussion of the aging of our own contemporary senses. In many ways the aging process affects our senses just as severely as Johnson's experiences. Yet the resiliency of the body means we can retain much functioning despite the cruel dictates of The Clock of Ages.

In this section, we are going to discuss how aging changes our five obvious senses: vision, hearing, smelling, tasting and touching. We will explore both the damage that has been done, the capabilities we retain and realize what an achievement it is, considering the complexity, that we retain anything at all. Thus it is quite a feat for a human to experience our senses in old age, even in the 20th century.

It was absolutely miraculous for a writer in the 18th.

The eyes have it

We shall begin our discussion with the sense of sight. As ever, to understand the process of aging on our eyes, we need briefly to discuss the relevant anatomy and physiology (Figure 28).

As you know, the eye is a sophisticated light-gathering structure. Its entire job is to focus light onto certain nerves. Those nerves become stimulated and send their excitation to the back of the brain. It is there that we 'see' our universe. To understand how this works – and how the years affect our sight, we will use the example of reading a book. Suppose you are examining a page from one of Samuel Johnson's commentaries on Shakespeare. We will follow the image of a word from the printed page to the back of your head.

You can see the word on the page because of the nature of the light in the room where you are reading. Made of photons, these particles/waves (no one has really decided what they are) are reflected off the page in a specific way. Our eye serves to collect and direct these particles/waves just like an antenna on a radio gathers radio signals.

The first structure the photons encounter is that fluid-filled bag on the

EYES

AND HUMAN VISION

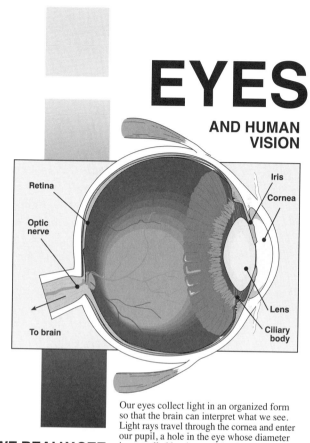

Retina

Optic nerve

To brain

Iris

Cornea

Lens

Ciliary body

Aging greatly affects the tissues in our eyes. To understand it, here's a brief review of our eye's anatomy.

WE REALLY SEE WITH OUR BRAINS

FIGURE 28

Our eyes collect light in an organized form so that the brain can interpret what we see. Light rays travel through the cornea and enter our pupil, a hole in the eye whose diameter is controlled by tiny muscles in the iris.

The light then enters our lens, whose job is to focus the light onto the back of our eyeball. To perform this task, the lens must be able to stretch and expand, a job controlled by muscle structures collectively known as the ciliary body.

When the light reaches the back of the eye, it is met by nerves that make up the retina. There are specific neurons which capture and respond to various characteristics (including wavelengths) of the received light. Once stimulated, the signal is sent through the optic nerve to the brain. The brain organizes and even helps interpret this input, allowing us to 'see' our world.

external surface of the eye, the cornea. As you know, the cornea is a transparent structure that guards the eye from most external objects in the room. Everything except light, of course. The photons reflected from the page are allowed to pass through this border patrol. Immediately they encounter not another structure, but a hole, the pupil. The number of photons let in is dictated by the size of this hole, which is controlled by the iris, a collection of muscles that created the pupil. The pigment of these muscles give us our eye color.

After passing through the pupil, the light enters the eye's interior and is about to be focused. This occurs through a structure that looks a lot like an onion, the human lens. It is made of many layers of clear protein and is also connected to muscles. These muscles, along with connective tissue and the blood vessels that feed them, are called the ciliary body.

Once the light rays are focused, they are sent through a jelly-like substance called the vitreous humor. This is mostly 'filler,' a space to journey through until the collected light hits the retina in the back of the eye. As you know, the retina is filled with photoreceptor neurons. These fire in salute whenever photons hit them. Such excitation is sensed by neighbor nerves, which throw the signal to yet other nerves. Eventually the signal is collected by a single stream of neurons, appropriately called the optic nerve. From there the signal is sent to the rear of the brain, where we perceive our visual world.

The aging of the eye

The eyes are remarkable organs in our body, and we could spend endless hours talking about the miracle of vision. As amazing as these tissues are, however, they are not immune to the aging process. Unlike the digestive system, nearly every part of the visual system is affected by the passing years. This causes us to lose certain functions as we age, ranging from a loss of color perception to a change in the way we relate to distant and near objects. We will talk about individual changes in the parts of the eye mentioned above, and then discuss how these changes affect vision (Figure 29).

The cornea, that fluid-filled bag which guards the entrance to the eye, ages as we get older. It is normally transparent, but certain changes in its molecular structure begin to cause a greater scattering of light rays with age. This change has a blurring effect on our vision. The overall shape of the cornea, in youth kidney-bean-shaped, also shifts. After the age of 60, the bag begins to flatten, which alters our visual field. To bring us back to youthful perceptions, we are required to buy corrective lenses.

The iris that governs the size of the hole, and thus the amount of photons permitted entrance, also undergoes age-related changes. Like so many muscles, the fibers atrophy as the years go by. Functionally, these changes tend to reduce the size of the pupil, allowing less light to come into the eye. That's why we need more illumination to read as we age. In addition, the iris becomes less flexible in our maturity. This means we do not adjust as quickly to sudden changes in the amount of environmental light.

Perhaps the structures whose changing has the most profound effects on our vision are the lens and the muscles that move it, the ciliary bodies. As you recall, the lens is a many-layered tissue. Through the years, more layers are added on to the outside, like the continuous addition of coats of paint. Unfortunately, the older inner layers are not first removed from the lens before the new ones comes on. This has the effect of compressing the older layers to the center and increasing the lens' diameter. By the age of 70, the the mass of the lens is almost tripled from its original value at age 20. This causes us to be more far-sighted, which is why we have to hold Samuel Johnson's books farther and farther from our eyes to see them. After the age of 70, there is a reversal of this trend and we become more near-sighted.

The effect of vision

The combined alterations of the cornea, iris, lens and ciliary bodies can account for most of the age-related changes in our visual perception. We don't see details as easily. That jelly-like substance between the lens and the photoreceptor neurons gets more opaque, further blurring our vision. Sudden bright lights from flashbulbs or oncoming cars cause our entire visual field to be disrupted – and we don't recover it as quickly once the stimulation passes.

Perhaps the most interesting aspect of our vision system's aging has to do with color perception. As we get older, the lens not only thickens, but takes on a yellowish hue. This reduces our ability to discriminate between colors in the green–blue–violet range. It also changes the way we see yellow. The net result is that our blues get darker and our yellows get

How vision changes with age

The aging process greatly affects our eyes, changing the way we see the world. Here's a summary of what happens to its various tissues.

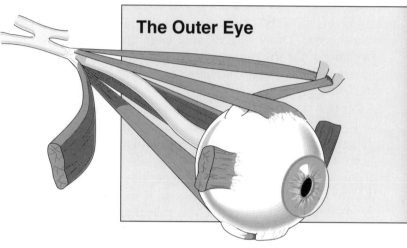

The Outer Eye

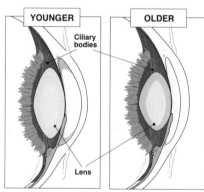

YOUNGER **OLDER**

Ciliary bodies

Lens

Aging of the human lens

As we age, new layers of tissue are added to the outside surfaces of the lens. The older tissue is not removed, however. The lens thus thickens with age, squeezing the older layers into the middle. This change causes us to become more far-sighted with age.

The ciliary body, the muscles, vessels and connective tissue also age. As the muscles atrophy, the ability to focus light rays to the back of the eye is diminished.

FIGURE 29

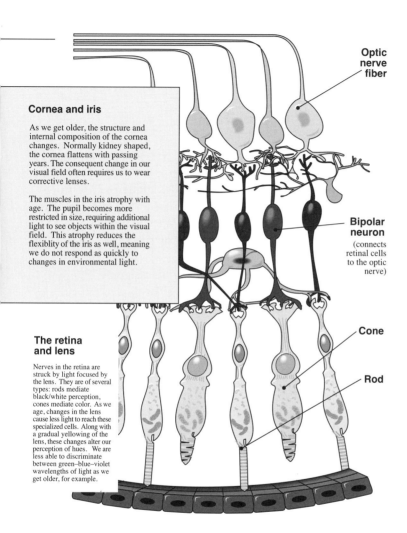

Optic nerve fiber

Cornea and iris

As we get older, the structure and internal composition of the cornea changes. Normally kidney shaped, the cornea flattens with passing years. The consequent change in our visual field often requires us to wear corrective lenses.

The muscles in the iris atrophy with age. The pupil becomes more restricted in size, requiring additional light to see objects within the visual field. This atrophy reduces the flexiblity of the iris as well, meaning we do not respond as quickly to changes in environmental light.

Bipolar neuron
(connects retinal cells to the optic nerve)

Cone

The retina and lens

Nerves in the retina are struck by light focused by the lens. They are of several types: rods mediate black/white perception, cones mediate color. As we age, changes in the lens cause less light to reach these specialized cells. Along with a gradual yellowing of the lens, these changes alter our perception of hues. We are less able to discriminate between green–blue–violet wavelengths of light as we get older, for example.

Rod

less 'bright.' We also see less violet. Some interesting studies have been done comparing palettes of painters who were creating art in younger and later years. Sure enough, as they age they use less dark blue and violet in their compositions, no longer having the ability to discriminate between them.

We are all hear

The Clock of Ages dictates more than just our perception of changes in our visual field as the years go by. Sweeping changes occur in our ability to perceive most kinds of input. This includes the ability to hear things. In this section, I'd like to talk a little about our auditory sensibilities and how they change in later adulthood (Figure 30). We will discuss of course a little anatomy before considering the effects of the years. And Irish lullabies.

This is a true story. The old woman awoke quite suddenly. She heard an Irish lullaby, sung by a strangely familiar voice. In the darkness of the night, she thought that a radio had suddenly been turned on, perhaps a wayward alarm clock. Annoyed, she turned on the lights, put on her slippers and searched around the house. Before she finished looking in the basement, she knew that the sound she had heard was not a radio. It was her grandmother's voice. Singing songs she knew as a child.

There were three melodies, sung in a constantly repeating, endless cycle. No one else could hear them and she thought she was losing her mind. Several visits to the doctor's allayed at least the mental fears. Something real and biological was occurring for sure, but it had nothing to do with mental illness. It had to do with a growing aneurysm.

Deep inside her head, a specialist told her, a blood vessel was on the verge of rupturing. It hadn't yet. It was merely bulging out, compressing brain tissue. It could be treated with medication, the doctor told her, and whatever else happened to the voices in her head, she needed to get on the medication. She did so, and the dangerous swelling very quickly went down. When it had reached a certain level, the voices in her head suddenly stopped.

What was occurring in the perception of this poor woman? The clinicians suggested that the swelling of the aneurysm did more than just compress nerves under her skull. They also stimulated them. The neurons that were firing may have held – or at least have had access to – certain

EARS

AND HUMAN HEARING

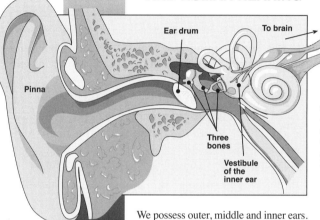

To understand how aging affects our hearing, we have to know something about its anatomy. Here's a brief review.

WE REALLY HAVE THREE EARS

FIGURE 30

We possess outer, middle and inner ears. Sound waves enter the outer ear through the large external flap called the pinna. The wave travels through the external auditory canal until it reaches the ear drum.

Attached to the drum are the bones of the middle ear, called the malleus, incus and stapes. When sound wave touches the drum, the ensuing vibrations also rattle these bones. The stapes is attached to a part of the inner ear known as the vestibule.

The inner ear is a complex structure filled with fluid and nerves. As the stapes vibrates, a sloshing motion is produced. This motion is registered by the nerves, which fire signals to the brain. In that manner, we hear sound.

aural memories of her childhood. When they were stimulated, she could hear them. And when the swelling went down, the nerves that triggered the melodic reverie were silenced.

For those who have ears

The events in this woman's head demonstrate in dramatic fashion the close association between head and ears in the perception of hearing. Just as eyes are the collecting apparati for light, ears are the collecting apparati for sound. But it is the brain where real hearing occurs, where nerves endlessly confront and organize the sound waves coming from the environment.

The brain has at its disposal three sections of sound-collecting tissues. These parts are termed the outer, middle and inner ears. The last set contains nerves that are wired to auditory centers in the brain. It is in these regions that you 'hear' sound.

The most obvious structure of the outer ear is the external flap, called the pinna. This bass-clef-shaped piece of skin works with sound the way a funnel works with liquid, focusing the flow into a narrow chamber. This chamber is called the external auditory canal. It ends at the ear drum (tympanic membrane), a delicate tissue that serves like a demilitarized zone between the outer and middle ears. When sound waves hit the tympanic membrane, it vibrates like the skin of a bass drum. This vibration can occur because of the support of muscles and ligaments.

The middle ear is the bridge between the outer and inner ears. This bridge is made of three bones, the tiniest ones in the human inventory. The first bone is connected to the ear drum, the third is superglued to the structure that makes up the inner ear. When the ear drum moves, the bones shake, producing a rattling that is transferred to this inner-ear structure. The job of the middle ear is sound conversion, transforming an energy that was moving through the atmosphere into energy that moves through solids.

That inner-ear structure has a name, of course. It is called the cochlea. This cochlea is stuffed with nerve endings that lead to the brain, and it looks something like a snail. It is also filled with fluid. The last bone (called the stapes, or stirrup) is connected to this cochlea, and when the bone rattles the cochlea feels it. How? You can think of the stirrup as a permanently attached plunger moving up and down in a stopped-up toilet bowl. As this bone moves back and forth in the ear, it sets up waves in the fluid of the cochlea. These waves can be sensed by the nerve endings inside the cochlea's curls. They become highly excited, depending upon the kind of wave they feel. This excitement is contagious, and soon other nerves in the auditory centers of the brain are stimulated, allowing you to hear things.

I sit and watch as ears go by

We will now examine what occurs in the outer, middle and inner ear as we age, and end with a brief description of how these changes affect our overall perception of the aural environment (Figure 31).

The changes in the outer ear are mostly cosmetic. As we age, the pinna loses some of its flexibility. As a result it begins to droop, becomes longer and wider and more filled with hair. This does not greatly affect our hearing.

Events that occur in the external auditory canal can, however. The sweat glands that normally keep the canal moist die off as the years go by. The effect is that the normal secretion of ear wax (called cerumen) becomes drier, crustier and less easily removed. Its build-up can dramatically affect hearing in older people, especially sound generated at low frequencies. Nearly one-third of hearing loss in older people occurs not because of some serious internal damage, but simply due to this build-up.

Changes also occur in the middle ear. As we age, the ear drum becomes thinner and more flaccid. The muscles that give it support also atrophy and the result is that the drum is less easily vibrated by sound waves. Those bones that connect it to the inner ear also change with increasing years. The joints that hold them together begin to calcify, stiffening the connections, causing them to be less easily vibrated. Although these changes sound as if they could result in great hearing deficits, the effect is less than you might suppose. In fact, the scientific literature is contradictory in exactly what these changes mean to our overall ability to hear. It can be safely said that hearing loss occurs, but exactly what effect the changes in the middle ear exert on this loss remains to be seen.

Structures in the inner ear are also affected with passing years. The nerves that sense the vibrations in the fluid, the spurious growth of nerve-impinging bone tissue, and loss of blood flow all contribute to a phenomenon known as, get ready for a big word here, presbycusis. There are

How hearing changes with age

Aging affects all three parts of the human ear, altering the way we hear our world. Here's what happens.

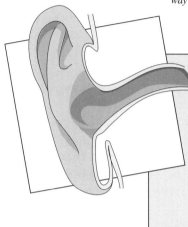

The Outer Ear
(the pinna)

Because of changes in certain proteins, the pinna begins to droop as we age. The external auditory canal grows hair. Due to a loss of sweat glands, the canal also gets drier. This leads to a crusty build-up of ear wax, which can severely affect hearing.

The ear drum becomes thinner and more flaccid with age. Atrophy occurs in muscles which normally support the membrane. As a result, the ear drum is less easily vibrated by sound waves.

AN IMPORTANT ASIDE TO THIS PAGE

The anatomical changes described here affect people's hearing differently. The ability to hear critical sounds can remain intact well into old age. There is a gradual loss of the ability to hear higher frequencies, starting around age 30. But the biological events described here do not mean severe hearing loss accompanies everyone into later adulthood.

Malleus

Incus

Stapes

Bones of the middle ear
Also called ossicles, these three bones are normally connected by joints, complete with ligaments and muscles.

As we age, these joints begin to degenerate. Ligaments and muscles follow normal aging patterns. Calcification occurs between the bones. These changes result in a less efficient transfer of vibrations from the ear drum to the inner ear.

FIGURE 31

The loss of specific frequencies

As we age, we lose our ability to hear specific frequencies (pitches) of sound. We generally retain the ability to hear the frequencies of the human voice. But other, quite specific losses across the frequency spectrum are a normal part of the aging process.

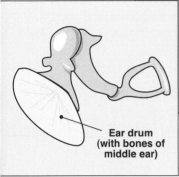

**Ear drum
(with bones of
middle ear)**

The Inner Ear

As we get older, the blood supply that nourishes the tissues in the inner ear decreases. Nerves that sense and conduct signals to the brain die and are not replaced. Excess bone formation occurs, the growth sometimes touching and impinging on auditory structures. These losses result in a type of hearing deficit known as presbycusis. There are several kinds of presbycusis, each kind leading to specific hearing problems.

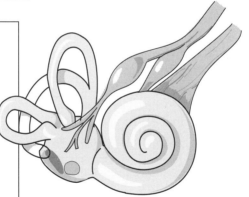

several kinds, but they all point to a single deficit – a loss of hearing at specific frequencies. The test that's usually given has to do with how loud a tone at a given frequency must be made in order for hearing to be perceived. This hearing loss usually starts around age 30 and continues well into the 80s. Usually the higher tones are the first to go.

As can be seen, nearly every part of the hearing apparatus is affected by the aging process. The deficits experienced in our vision and hearing are so consistent that they have almost become clichés for the portrayal of old age in the media. Fortunately, this depressing fact is not true of all our senses. To describe an apparatus that has an amazing resiliency throughout our maturity, I would like to shift our focus from the sides of our head to the inside of our mouth. We will next consider the effects of aging on our ability to taste.

Days of wine and roses

I once saw an old man whose hearing was greatly impaired. He was giving a lecture on wine and was greatly revered, not for his acoustical talents (I could barely understand him), but because of some gustatorial ones. The man was a French wine taster, and he had the distinct privilege of tasting some of the world's great beverages.

He related the story of one of the world's great losses. A bottle of Château-Lafite claret was put on the auction block at Christies in London. It was unique not only because of its age, but also because of its previous owner. The bottle used to belong to the wine cellar of Thomas Jefferson. It even carried the famous US President's initials. Bidding stopped at $157 000, and the bottle was awarded to the American millionaire Christopher Forbes.

'The problem and tragedy,' the wine taster said in his broken English, 'had to do with the lights in the exhibition hall. During the auction the lights dried the cork out, thus shrinking its diameter. Ten months later the cork was sufficiently small that it fell through the bottle, rendering the wine undrinkable. It wasn't discovered until later, but his investment, and one of the world's great wines, was destroyed.'

A matter of taste

This lecture illustrated for me the dramatic discrimination The Clock of Ages exerts on our senses. The gentleman could barely hear the questions following his interesting story, but his palate was so finely tuned that it was still winning awards. It points to the fact that our ability to taste things stays in our sensual repertoire far longer than might be expected. In this section we are going to discuss this remarkable sense, first by exploring how we taste things, like a good bottle of wine, and then by discussing how the sense ages (Figure 32).

We experience the tastiness of foods and drinks because of an interaction between taste buds and nerves in the tongue (our sense of smell also gets into the act). A taste bud is a collection of cells that peak through a hole in the tongue, like a cork peaking out of the top of a wine bottle. They are connected by a nerve underneath them which runs off to the brain.

But only the top tip of the cell, which has a hair on it, is exposed to the environment. This tip, which is part of the cell's outer membrane, also possesses receptors of specific shape on its surface. These receptors cause us to experience taste. When we eat something, the saliva-drenched food particle has to touch these various receptor shapes. If the food finds a receptor into which it can fit, like a key into a lock, the food particle will bind. This binding excites the nerve underneath and we experience the sensation as taste.

As you may know, different regions of the tongue are outfitted with different types of taste buds – all assigned certain sensitivities. There are salty regions, which are stimulated by salt molecules (no surprise there). There are sweet regions, which respond to alcohols, certain chopped up proteins, sugars and even certain salts. There are sour regions, which respond to acid. There are even bitter regions on our tongue, which respond to groups of molecules such as alkaloids.

How it ages

There is some controversy surrounding our ability to experience taste as we age. This is true for two reasons: (1) our perception of taste is intimately associated with other senses, like smell, appearance and texture; and (2) taste buds regenerate. Cells within the taste bud die about every ten days and are fully replaced. Even if the nerve is destroyed, other buds will form around a new nerve growing towards the surface of the tongue. The net result? We lose very few taste buds with age.

These facts are contrasted with a large body of data that says we lose our taste sensitivity as we age. This means that it takes more molecules

The aging of taste

Our ability to taste our environment changes as we get older. Here's how taste works in our bodies and what happens when we age.

HOW WE TASTE THINGS
We experience taste through a collection of cells connected to a nerve. This structure is called a taste bud.

Various regions of the tongue are responsible for specific tastes, as illustrated on the right. Flavors are experienced because a molecule of food lands on receptors on the taste bud's surface. This excites the nerve beneath it. The subsequent signal is shuttled to the brain.

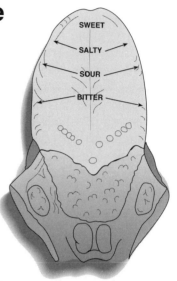

THE CURIOUS CASE OF BABIES

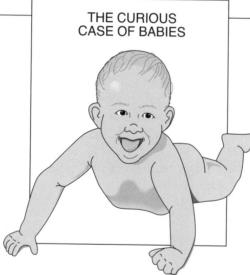

Taste buds die and are rapidly replaced. We thus lose our sense of flavors only gradually with age, with equal reduction of all areas.

The loss isn't quite as steady in early life. When we were born, taste buds literally covered our mouths. There were flavor sensors on the roof of our mouths, in our throats and over the lateral surface of our tongues. They died off quickly, however. By the time we were 10, most of these extra taste buds were gone.

FIGURE 32

of a certain substance on our tongue for us to recognize the flavor later in adulthood. The change in sensitivity appears to be true for most taste experiences, whether salty, sweet, sour or bitter. As a result, we tend to enjoy food less as the years advance.

Why this decrease should occur unassociated with taste bud loss is not known. It has been explained in many ways, even to the point of questioning the data on sensitivity loss. Until more research is done, the discrepancy between biology and perception will remain. There *is* some hope for a solution, however. The powerful techniques of molecular biology have cloned many of the genes involved in taste perception. As we understand more about the process in molecular terms, this discrepancy between taste buds and perception will be better understood.

For whom the smell tolls

Our ability to taste the many foods of our environment, as stated earlier, is intimately connected to our ability to smell. To observe this linkage, simply try to taste food when your nose is stopped up by a cold. The taste almost seems to disappear. This smelling talent, called olfaction, is probably one of our least developed senses. And the intimate link between it and our taste buds makes this area of human biology rather difficult to study.

Even if our sense of smell doesn't compare to that of many creatures in the animal world (bloodhounds have noses one thousand times more sensitive to most odorants, for example), the tissues in our nose still perform vital functions. We will first describe what this organ looks like and then talk about how it ages (Figure 33).

At the top part of our nose, just behind the area between our eyes, lies the organ that allows us to smell things. It is called the olfactory epithelium. This tissue is filled with cells that have hairs on them and are plugged into nerves, just like our taste buds. These nerves, when stimulated, send a signal to the brain and we perceive smells.

What these cells and nerves sense are specific molecules, termed odorants, in the air. These odorants go up the nose every time we breathe and look for a place to which they can bind. In the olfactory epithelium, they find such a place. The surface of cells in the olfactory epithelium possess receptors of varying shapes. Like their counterparts on the tongue, these receptors can bind to specifically shaped odorants in that lock and

SMELL ANATOMY AND FUNCTION

Intimately connected to our ability to taste, the sense of smell is one of our oldest senses. Here's how it works and how it ages.

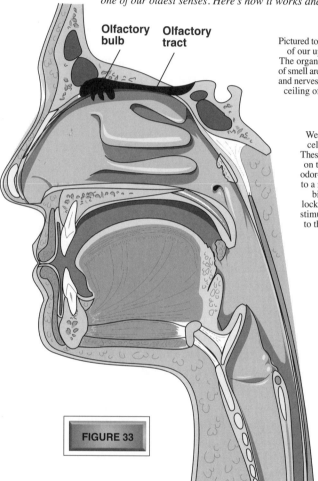

Olfactory bulb

Olfactory tract

Pictured to the left is a cross-section of our upper respiratory system. The organs that mediate our sense of smell are the olfactory bulb, tracts and nerves. They are located on the ceiling of the cavity, as shown in the drawing.

We identify odors with the cells in the olfactory bulb. These cells possess receptors on their surfaces. When an odor-bearing molecule binds to a receptor (it is a specific binding, like a key into a lock), the olfactory nerve is stimulated. The signal goes to the brain and we register the smell.

Exactly how or if this sense ages is not known. Our ability to discriminate odors remains fairly constant through age 65. After this time there is some loss, but the amount of reduction varies widely between individuals. One reason for this difficulty is that environmental pollutants can play a large role in affecting the integrity of the olfactory system.

FIGURE 33

key model we discussed earlier. When this odorant-to-receptor binding occurs on the cell surface, the nerve below it is stimulated. The nerve then sends a signal directly to the brain, where the smell is perceived. Different odorants bind to different receptors, or even combinations of receptors, allowing different types of smells to be experienced.

How does our sense of smell age?

The controversies that surround our aging taste sensations also surround our aging olfactory capabilities. One of the great difficulties is trying to classify various odorants and their sensations (is this a strong smell? a weak smell?), and creating quantifiable tests that eliminate the bias of experience. Another has to do with the environment, and its potential effect on the delicate tissues within the nose. Have you been around a polluted area most of your life? Do you smoke? Were you around someone who did?

Despite these obstacles, some facts are starting to come into focus. The most interesting one is that our olfactory capabilities hold up pretty well in later adulthood. In tests involving the ability to detect and distinguish between odors such as mint, coffee and anise, older adults discriminate just as well as teenagers. Only when we reach the age of 65 is there a noticeable change. And even here, the extent of the deterioration varies widely between experimental subjects.

The genes that code for the receptors responsible for olfactory perceptions, like those in the tongue, have been isolated. As we understand more about the combinations necessary to produce sensation and their regulation in late adulthood, what really occurs in the nose will be better understood. Then we will be able to relate ambiguous field tests with solid biochemistry – and watch how these change as the years go by.

The Clock of Ages appears to vary widely in its effects on our senses. As we have just seen, taste and smell appear to be minimally affected, especially when we consider what happens to sight and hearing. This seeming capriciousness is also observed in the last sense we will discuss in this section, the perception of physical touching. Like tasting and smelling, this group of sensations is amongst the hardest of all our experiences to measure accurately. As a result, understanding the effects of the passing years on our sensitivities to events like pressure, pain, cold and hot perceptions can be quite difficult. Especially considering how important touch is to the human experience.

A touching reminder

It is intriguing to see a human baby begin to relate his or her senses to the rest of the world. One of the first sensations a baby experiences is not the sense of smell, but the sense of touch. It appears to be important in development and not just for humans. New-born animals with skin that is stimulated through licking, for example, grow up more resistant to bacterial infection than those without. They gain more weight and are more active throughout life.

Touching affects humans too, psychologically as well as physically. Babies that are held, fondled and caressed are much less likely to suffer depression, weight loss and disturbed sleep. Like their animal counterparts, they also have greater resistance to infection. It is obvious that touch, one of the first senses we acquire, has a great deal to do with the quality of life.

The biology of touch

It is one thing to say that touch is important. It is another thing to quantify the sensations. We can categorize them easily enough, but further classification makes understanding their biology very difficult indeed.

We basically respond to three types of touch:

(1) Pressure;
(2) Pain;
(3) Heat/cold (thermal sensitivity).

We'll go through each type of sensation and describe, where possible, how the aging process affects the biology.

Pressure When someone touches our skin, our brain is notified through a series of receptors. These receptors have odd names, like Meissner's corpuscles, Merkel's disks, Rufinni end-organs and pacinian corpuscles. There are variations in their occupational descriptions, but the end task is similar. Their job is to alert connecting nerves that pressure is being applied.

As we age, many of these receptors die off. Even pacinian receptors, which can regenerate fairly easily, are reduced in numbers as the years go by. As a result, there are changes in our ability to perceive pressure. The skin in our hands undergoes a decreased sensitivity to touch. After age 50, our ability to perceive vibrations in the lower part of our body is diminished. The loss is further compounded by damage in the nerves that conduct pressure information signals to the brain.

Pain The sensation of pain is mediated by free nerve endings located in most of our body's tissues. There are many kinds of pain that can be perceived, including mechanical (pulling a ligament, for example), pressure (a puncture) or thermal (burns, from either hot or cold sources).

The problem with understanding pain has to do with the enormous variations in its perception. And tolerance. Some individuals are more aware of their bodily sensations than others. There are those who try to minimize any sensations of physical pain and others who holler at a hangnail. There are gender and cultural considerations as well. As a result, researchers in the area have to wade through a thicket of psychological factors in order to reach physiological conclusions. There is even such a thing as a 'false affirmative' rate (reports of pain when no stimulus has been applied in an experimental situation) created in an attempt to separate subjective experiences from physical phenomena. Regarding age-related changes, investigators have not been very successful in separating the subjective from the objective.

Nonetheless, scientists have attempted responsible studies in this area. And there is compelling reason to do so. As we age, more painful experiences per unit time could potentially occur; joints wear out, bones break, organs become less efficient, disease processes go unchecked.

But potential does not mean actual and what those changes mean in terms of perceptions is quite unknown. For example, there are some data that older individuals require more intense pain experiences in order to perceive that something is wrong. This may be due to a decrease in the number of pain receptors throughout the body, in which case the overall experience of pain may not match the physical deterioration. As much sense as this might make, it has not yet been proved experimentally. There is much work to be done regarding how pain is perceived in our later years.

Heat/cold (thermal sensitivity) If pain responses are not well established, whether physical or mental, the situation is even worse regarding our ability to detect hot and cold. The reason is that even the source structures mediating this perception in our skin are not well understood. We know that there are probably receptors for warmth and other receptors for cold. The receptors are probably groups of nerves, with free ends that are just waiting to detect a thermal difference. There is some evidence that the receptors that detect cold outnumber the receptors that detect warmth.

Measurements of thermal sensitivity are complicated by our perceptions. Specifically, we have a curious psychological ability to adapt to temperature differences. This adaptability plays such a large role in our perceptions that what we experience as cold in one season may be hot in another. The more we adapt, the more temperature difference we will need to experience in order to perceive a change. We cannot detect temperature alterations (within a narrow range) if they occur gradually. Sudden thermal events trigger not only hot or cold sensations, but also pain. As if the picture needed further ambiguity, the problems associated with understanding pain are brought into this research as well.

Is there anything we can say about aging and thermal sensitivity? There is some research indicating that higher thresholds of heat are needed before older adults can experience warmth, but until the biology of human thermal sensitivity is better understood, the data regarding age-related changes in detecting hot and cold are mostly commentary.

To summarize the sense of touch in the skin The best thing we can say about the aging of the sense of touch, in all its many facets, is the following: something happens. We know just enough physiology to understand that some kind of change occurs as the years go by. We just aren't smart enough yet to discover the role perception and biology have to play in the overall experience.

How to make sense of this

One only need review the life of Samuel Johnson to understand what it means to lose in piecemeal fashion various human senses. For us in the 20th century, the aging process exerts a far gentler force on our senses than Johnson experienced. Our eyes don't suddenly go blind, but gradually lose specific capabilities that take decades to be felt. The same is true of our hearing. Our sense of taste and smell don't appreciably lose anything at all. We don't truly understand what really occurs to our sense of touch. Despite these changes, we retain a tremendous ability to examine, interpret and affect our environment. And this we can do even past the age of Samuel Johnson's death.

As ever, the question of why this has to happen at all is very important. We end up by confronting some very familiar mysteries. What governs the changes in our sense organs at the cellular level? Why do iris muscles have to atrophy? Why must the bones in our middle ears become less flexible? If natural selection becomes untethered at the point where we lose reproductive function, what genes are ignited that still keep us changing with passing years?

The curious thing about aging is that it affects our reproductive capabilities, even though we say aging is untethered from normal selective forces. What that means in biological terms is very important, and I have saved its explanation until last. If you'll read on, we shall discuss this final, most intimate interference of The Clock of Ages.

11

The aging of the reproductive system

These words could only come from the mind of the world's most famous seducer: 'Life is a wench that one loves, to whom we allow any condition in the world, so long as she does not leave us.'

The comment was penned in the last months of the life of Giovanni Casanova, the legendary lover of the 18th century. Fittingly, he met his end after no fewer than 11 bouts with venereal disease, and died because of complications from one of them. But he paid for his dalliances in more than just mortality. Because of these ailments, his amorous career actually stopped 13 years before his life expired. He spent those years eating food in the kitchens of European nobility, causing one biographer to quip: 'Since he could no longer be a god in the gardens, he became a wolf at the table.'

The father of countless illegitimate children was himself born a bastard in the Venice of 1725. Although he never fully established his paternity, he did know his mother, a famous actress of the day named Zanetta Farusi. He grew up in her household, half-brother to a number of other children whose paternity could never be fully established, either.

Casanova's first encounter with the sensuous life occurred at age 11, in the experienced arms of the woman who normally gave him a bath. His teenaged years were full of heterosexual explorations, where he fine-tuned his history-making skills of seduction. But his early career also included men. Casanova entered a seminary and was promptly kicked out after a homosexual tryst with one of the priests.

The rest of Casanova's sexual history – at least through age 49 – comes to us via his massive 4545-page autobiography. He recounts in its pages several lifetimes of sexual intrigue. Not committed to long-term emotional relationships, Casanova rather liked to break them apart. He thought of marriage the way one thinks of graveyards, at one point calling the institution the 'tomb of love.' The freedom to enjoy multiple partners was more to his taste – and the more illicit, the better. He liked the fine art of adultery, for example, describing his adventures as 'the inexpressible charm of stolen pleasures.'

Casanova was rather less enamored of the microbiological damage his liberties produced. Assisted by an equally dangerous treatment (the toxic heavy metal, mercury), his diseases permanently weakened his urogenital tract. He became sexually inactive by age 60 and eventually acquired prostatitis, the disease that would kill him. In his dying hours, he called a priest and confessed as many of his illicit love affairs as possible. His last words: 'I die as a Christian.'

Casanova's life expired the next morning, on June 4, 1798. He was buried in a 'lost' grave whose tombstone was only uncovered in 1922. No trace of his body has ever been found.

This section

Though one might quarrel with his emotional allergy to commitment, Giovanni Casanova's amorous vigor brings up an important point about sexuality in later life: if left untroubled by venereal disease, our sex life can exist in later life with great physical enthusiasm. It is perhaps an odd juxtaposition, because our ability to have children declines with age; in menopause there is a dramatic cessation of the capacity altogether. Thus we are in the unique position of being able to use our sexuality for purposes other than procreation of the species.

The Clock of Ages has a great deal to say about the future of reproductive tissues in both genders. It is the purpose of this section to describe some of those changes. We will begin by discussing some of the anatomy and aging of the female reproductive system, and end by explaining what happens to males.

What happens to women

The anatomy

When one thinks of age-related changes in female reproductive potential, the word 'menopause' comes to mind. It's also called 'the change of life' or 'the climacteric.' All the terms refer to the halting of the menstrual cycle in middle-aged women. The word menopause itself is a collision of two languages, the Latin *mensis* meaning month and the English 'pause' meaning exactly that – pause. But the change in the menstrual cycle is permanent, unless interfered with chemically. When it shuts down, a woman's reproductive history is finished. The great engines of natural selection grind slowly to a halt, with lots of consequences to the female body.

To understand the signals that tell the menstrual cycle to quit, we must first understand the chemistry that keeps it going. There are many hormones involved, all with abbreviations that would make the American Defense Department jealous. For example, there is FSH (follicle stimulating hormone) and there is LH (luteinizing hormone). These chemicals act like field generals, sent down from command central in the pituitary gland. FSH stimulates the production of the multi-functional hormone estrogen. This hormone controls the development and maintenance of the female reproductive structures and secondary sex characteristics (Figure 34). LH helps stimulate that other legendary hormone progesterone, which helps supervise a pregnancy once fertilization occurs.

How the system ages

These complex hormonal interactions begin to change between the ages of 45 and 50. As dramatic as the change becomes, the precise molecular signals The Clock of Ages exerts over women are not fully understood. It is known that the ovaries are told to quit making a chemical called estradiol. This chemical is *the* most powerful estrogen. Why does it stop? Remember that FSH sent from the brain gives the follicles certain instructions like 'make estrogens.' In menopause the follicles stop obeying orders. The amount of estrogen, especially this estradiol, becomes very scarce. What makes the follicles so recalcitrant in later years? Nobody knows for sure.

Even considering the temperamentality of a given follicle, menopause does not happen suddenly. The menstrual cycle slowly gives clues that a change is afoot, signalling as early as ten years prior to menstrual cessation. One of the most noticeable clues is the reduction in the number of days it takes to go through a period. At age 30, the normal time is 28–30 days. By age 40, the time has shrunk to 25 days and several years

The role of estrogen in female anatomy

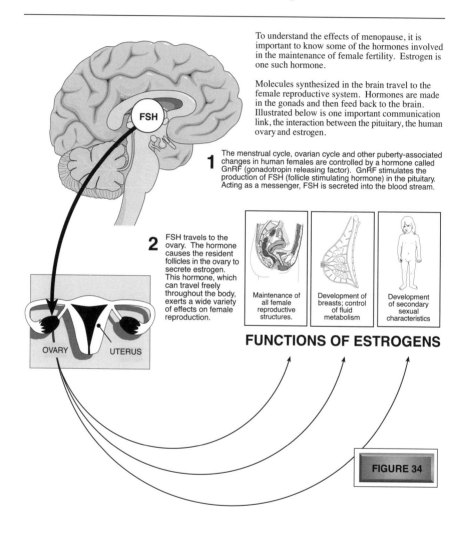

To understand the effects of menopause, it is important to know some of the hormones involved in the maintenance of female fertility. Estrogen is one such hormone.

Molecules synthesized in the brain travel to the female reproductive system. Hormones are made in the gonads and then feed back to the brain. Illustrated below is one important communication link, the interaction between the pituitary, the human ovary and estrogen.

1 The menstrual cycle, ovarian cycle and other puberty-associated changes in human females are controlled by a hormone called GnRF (gonadotropin releasing factor). GnRF stimulates the production of FSH (follicle stimulating hormone) in the pituitary. Acting as a messenger, FSH is secreted into the blood stream.

2 FSH travels to the ovary. The hormone causes the resident follicles in the ovary to secrete estrogen. This hormone, which can travel freely throughout the body, exerts a wide variety of effects on female reproduction.

Maintenance of all female reproductive structures.

Development of breasts; control of fluid metabolism

Development of secondary sexual characteristics

FUNCTIONS OF ESTROGENS

FIGURE 34

later, to 23 days. Irregularities also increase. Many times the body will undergo an ovarian cycle when no ovulation occurs. The quality of the egg produced also changes. After the age of 35, the ova are more 'defective' genetically. If fertilization occurs, the babies produced are more likely to have birth anomalies.

A many-faceted event

Estrogen is involved in many aspects of female reproduction. As a result, the echoes of its loss are felt throughout the body. One of the most dramatic effects has to do with physical appearance. As estrogen is greatly involved in the upkeep of secondary sexual characteristics, its loss changes the maintenance schedule of these tissues.

This change in supporting processes can be seen in the structure of the female breasts. In later adulthood, they begin to sag. As drooping is a normal consequence of the mammary's sojourn through a gravitational field, that's not necessarily surprising. With the loss of estrogen, however, mammary gland tissue is more rapidly replaced by fat. Because fat is not as strong, the sagging is aggravated. As elastin and collagen fibers are changed, increased wrinkling occurs, too.

Other structures in the breast are affected after menopause. The nipples decrease in size, as does the surrounding areolar tissue. These structures are not as easily made erect when externally stimulated. Stretch marks in the breast, obtained through either pregnancy or the normal swelling and shrinking during the menstrual cycle, get darker with age as well.

Throughout the body, subcutaneous fat begins to accumulate in the torso, especially near the waist. This can also be seen in the skin of the neck, the arms and thighs. The net effect of such accumulation is that uneven bulges are created, dramatically changing the appearance of a woman's figure. This redistribution of fat can also be seen on the face. As we age, the fat begins to migrate away, creating a hollower visage. Coupled with wrinkling, drying and thinning of the skin, this loss of estrogen greatly affects the physical appearance.

While the effects of estrogen are dramatic, other hormonal changes occur during menopause too, many of these only now being addressed in the research laboratory. In fact, menopause is best described as a multifactorial process, involving millions of cells and enormous changes in gene expression. The entire process can be likened to a symphony orchestra,

with various members suddenly switching musical scores during the performance. While sound can still be heard, a dramatic change occurs in
the way we hear the themes (Figure 35).

What does it do to sexual activity?

Whereas these physical changes in human biology are fairly well documented, changes that aging exerts on our normal arousals and attitudes
towards sexuality are less established. This paucity of information is
unfortunate, because human sexuality has significant psychological components that must be dialed into the equation when we consider aging
and sex. One of the great questions women and men ask about menopause
is: what does this do to a woman's sexual activity and responsiveness?
The question can be answered in several ways, first by describing physical
changes and then by discussing psychological components.

There are physical changes in the organs that mediate sexual response
in the female genitalia. The pubic hair becomes sparse and coarser. The
labia majora and minora become more wrinkled and thinner, primarily
due to that fat migration we discussed earlier. There is general atrophy
of the three layers of epidermis surrounding the vulva, which is part of
the skin's normal aging process.

There are also changes that occur in the vagina. The cells that make
up its walls become weaker and the structure is more susceptible to tearing. The vagina also becomes drier, and along with other changes in the
internal microenvironment, is rendered more susceptible to infection.
With a loss of elastic tissue, the vagina is also less capable of shrinking
and expanding; it thus becomes less accommodating to the insertion of a
penis.

But does all this have to do with a physical change in sexual response?
The answer is yes. Note that this question was *not* 'Is there a change in
sexual interest?' As stated before and as we'll see in a moment, sexual
interest does not necessarily wane with age. But the anatomical changes
mentioned above exert a biochemical change on certain sexual responses.
These changes are outlined below.

As you probably know, sexual response can be divided into four phases:
excitement, plateau, orgasm and resolution. Here are data from studies
that compared sexual responses in early adulthood (women aged 20–40)
with later adulthood (women aged 50–78).

Excitement In early adulthood, the time needed for vaginal lubrication after sexual arousal is between 15 and 30 seconds. In later adulthood, that time increases to 1–5 minutes.

Plateau In early adulthood, the vagina can expand essentially without pain during arousal. In later adulthood, the upper limit for expansion is reduced. The minor labia are reddened because of increased blood flow in younger adults. In older adults, the reddening does not occur. The clitoris elevates and flattens against the body in younger women. This does not occur in later adulthood.

Orgasm In younger adults the vagina contracts and expands in smooth and rhythmic waves, usually 8–12 contractions in 0.8-second intervals. In older adults, the 0.8-second interval is retained, but the overall number is reduced to 4–5 contractions. The uterus also contracts during orgasm, and in older adults this is sometimes painful.

Resolution The time it takes to return to a pre-arousal state is longer in younger women, more rapid in later adulthood.

Sexual interest

Do these alterations in physical responses correlate with changes in sexual interest? The answer is a happy and enthusiastic 'NO!' In fact, one study demonstrated an increase in sexual interest and consequent activity in married women between the ages of 69 and 76.

As is true with so much of human sexuality, the level of interest appears to have many more psychological components than physiological ones. When sexual interest declines, it appears to be more related to the presence and/or attitude of the woman's partner. Unmarried women in later adulthood tend to have less sexual interest than married women. If the husbands lack interest, either for psychological reasons or because of poorer health, the woman tends to lose interest also. The woman's perception of her desirability also greatly influences her sexual interest.

Thus it appears that the aging process in women affects reproductivities but not proclivities. Intimate changes of such a dramatic nature have profound psychological consequences, but none of these need be reflected in overall sexual attitudes and enthusiasms.

The changes of menopause

The alterations occurring in menopause are felt throughout the body. How the outer female body, genitals and experience of sex change are outlined here.

1 FACIAL CHANGES

Fat begins to migrate away from facial structures after menopause. This is coupled with a loss of water. As a result, skin on the cheeks, under the eyes and around the jawbone begins to sag. A hollower visage is created.

2 BREAST CHANGES

Menopause accelerates the sagging of breasts. With the loss of estrogen, mammary gland tissue is more rapidly replaced with fat. Because fat is not as strong, the sagging is aggravated. The nipples decrease in size, as does the surrounding areolar tissue. Breasts do not become as easily erect when externally stimulated. Existing stretch marks darken markedly after menopause.

3 PLACEMENT OF FAT

The distribution of subcutaneous fat (tissue beneath the skin) seen in the face also occurs throughout the body. Fat starts to accumulate in the torso, especially near the waist. This accumulation can also be seen in the skin of the neck, the arms and thighs. Uneven bulges are created, dramatically altering the appearance of a woman's figure.

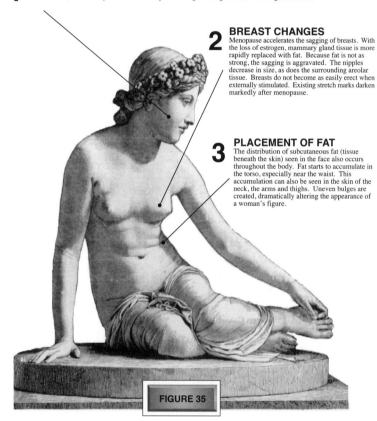

FIGURE 35

CHANGES IN GENITALIA

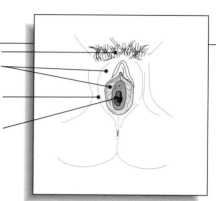

The pubic hair thins and becomes coarser.

The labia majora and minora become more wrinkled and thinner, primarily due to that fat migration we discussed on the left.

There is general atrophy of the three layers of epidermis surrounding the vulva, which is part of the skin's normal aging process.

The cells that make up the vaginal walls become thinner and the structure is more susceptible to tearing. The vagina also becomes drier, and along with other changes in the internal microenvironment, is rendered more susceptible to infection. With a loss of elastic tissue, the vagina is also less capable of shrinking and expanding.

CHANGES IN SEXUAL RESPONSE

Female sexual response has been divided into four phases: excitement, plateau, orgasm and resolution. Summarized below are changes that occur in each phase after menopause.

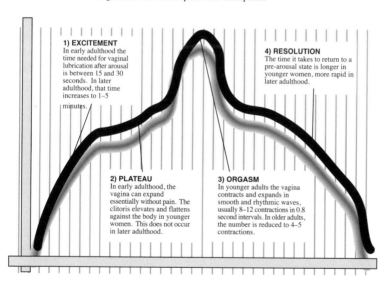

1) EXCITEMENT
In early adulthood the time needed for vaginal lubrication after arousal is between 15 and 30 seconds. In later adulthood, that time increases to 1–5 minutes.

4) RESOLUTION
The time it takes to return to a pre-arousal state is longer in younger women, more rapid in later adulthood.

2) PLATEAU
In early adulthood, the vagina can expand essentially without pain. The clitoris elevates and flattens against the body in younger women. This does not occur in later adulthood.

3) ORGASM
In younger adults the vagina contracts and expands in smooth and rhythmic waves, usually 8–12 contractions in 0.8 second intervals. In older adults, the number is reduced to 4–5 contractions.

But why stop reproduction at all? What does natural selection gain by altering estrogen production in middle-aged women? The answer is that nobody knows. One always has to be careful whenever biology affects an organism's reproductive output. Since natural selection is bent on survival into the next generation, when that productivity is altered, the rules begin to change. If selective pressures really are removed from us once we lose reproductive fitness, the capricious genetic aftershocks could affect any process. We can only be thankful that it leaves our libido alone.

What happens to men

The changes observed in women are often judged as quite dramatic when compared to changes in men, simply because menopause is so conspicuous. As the years go by, men also lose their reproductive capacity, although this loss is more like a reduction than a cessation. The change in fertility is mediated primarily by a reduction in the amount of viable sperm available per ejaculation. There are also changes in semen production and the ability to produce and maintain an erection. Even testosterone may be affected. We shall examine each of these aspects of male reproduction and conclude with a comment on sexual activity (Figure 36).

What happens to testicles

As you know, human testicles (testes) have two primary outputs: sperm and testosterone. These testes look as if they have been stuffed with cooked spaghetti noodles. They are filled with long tubes, called seminiferous tubules, which are the sites of sperm production. It's quite a manufacturing assembly line; if those tubes were stretched out, they would create a string three feet long.

The testes are ordered to make a batch of sperm from the same organ a female uses in her reproductive cycle, the pituitary gland. And males use identical hormonal messengers, FSH molecules, to deliver the manufacturing instructions. Once FSH arrives, new sperms are made in an assembly line fashion via the help of an extraordinary cell called a Sertoli cell. Development proceeds from the outside tubes to the inside, with the final product dumped into a structure known as the epididymis.

As the sperm turns

Beginning around age 50, there is a breakdown in the quality control of sperm manufacture. As men age, the seminiferous tubules begin to narrow. This is due to thickening of connective tissue within the tube itself. Such a change won't necessarily cause pain, but it does render the tubule increasingly non-functional. The tubes can deteriorate to the point of collapse. As you might expect, sperm production is dramatically altered.

When such alterations occur, the command centers in the pituitary gland sense that something is wrong. In response, they send more hormones laden with increased production orders (like FSH) to the reproductive organs. The testes ignore these new instructions and continue to supply lower amounts. In some ways, it is reminiscent of the female follicle's refusal to follow FSH's instructions. Unlike in females, however, there is never a cessation of viability. Indeed, the oldest verifiable father was 94 at the birth of his last child.

Semen production

The ability to make semen also changes with age. As you know, new sperm are supplied with semen, which is a sticky and proteinaceous substance, prior to ejaculation. This gummy fluid gets its viscosity from having so much sugar (fructose) in it. Semen is produced by the prostate gland, which stores the stuff until needed for ejaculation. Like the testes, this organ also possesses tubes and sugar-producing cells. It even has some smooth muscle.

After the age of 40, anatomical changes begin to occur in the prostate gland. Certain tissues (especially in the back of the gland) begin to atrophy. By age 60, there is a dramatic reduction in the prostate's ability to make semen due to this accumulated deterioration. As the smooth muscle degenerates, it is replaced by inelastic connective tissue. Hard masses can even appear in the organ. The result? Reduced volume and reduced pressure of semen per ejaculatory event.

A common age-related problem in men occurs with a change in size of the prostate. The exact cause of the problem is not known, but in many males the gland cells and the connective tissue in the middle of the prostate start overgrowing – as if they'd been given fertilizer. This

Aging and male sexuality

Changes in male reproduction include alterations in sperm count, prostate size, erections and the experience of sex. Here's a summary of what happens.

MALE ANATOMY

Sperm are made in the testicle and stored in the epididymis (see below). Prior to ejaculation, the sperm are mixed with semen from the prostate. Shown in the box on the right are cross-sections of these structures. An enlargement illustrating the path of the sperm (arrows) is drawn below.

MALE REPRODUCTIVE GLANDS IN CROSS SECTION

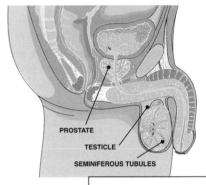

PROSTATE

TESTICLE

SEMINIFEROUS TUBULES

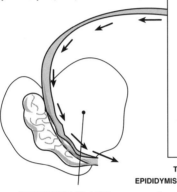

TESTICLE

EPIDIDYMIS

PROSTATE GLAND

Semen is produced by the prostate gland, shown here in outline form. A combination of sugars, proteins and water, semen is stored in the prostate until needed for ejaculation. Like the testes, this organ possesses tubes. The prostate also possesses sugar-producing cells and is lined with smooth muscle.

Due to accumulated deterioration, there is a dramatic reduction in the prostate's ability to make semen after the age of 60. Hard masses can even appear in the organ. A reduction in volume and semen pressure per ejaculatory event occur in later adulthood.

A common age-related problem in men is a change in size of the prostate. Certain tissues within the gland begin accelerated growth. There is increased synthesis of connective tissue. This increase in tissue presses on the tubes that normally help void urine. It can be increasingly difficult to urinate as a result.

SPERM COUNT

The testicles manufacture sperm in an assembly line fashion. The final product is stored in the epididymis prior to ejaculation. After age 50, there is a breakdown in the quality control of sperm manufacture. Due to a thickening of tissue within, the seminiferous tubules begin to narrow. Eventually, they deteriorate to the point of collapse. The resulting sperm production is dramatically altered. That's why in later life, men's sperm count is lower than in early adulthood.

The pituitary senses this loss and sends more testes-stimulating follicle-stimulating-hormone (FSH) to increase production. The testes are unable to respond and the sperm count remains low. There is never a cessation of viability, however. Males can father children well into their 9th decade of life.

FIGURE 36

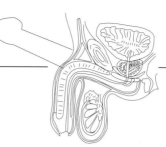

ERECTIONS

Erections occur because of increased blood flow into the penis. With age, there is greater difficulty in producing and maintaining erections. This occurs because the blood vessels become more rigid, allowing less fluid to flow through them. There is increased growth of connective tissue on the inner surfaces of the penis, rendering it less flexible.

SEXUAL RESPONSE

Comparative studies on the phases of male physiology during sex have been done on younger and older males. Listed below are some of the results.

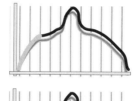

EXCITEMENT

The time needed to obtain an erection after stimulation was 3–5 seconds in the younger sample and 10 seconds to several minutes in the older population.

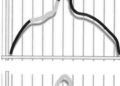

PLATEAU

Quick development of pressure for ejaculation was experienced in the younger population. The older population took longer to feel the pressure. Psychologically, those in later adulthood felt less of a need to ejaculate during a sexual episode or even across several episodes.

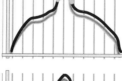

ORGASM

The penile urethra (the tube in the center of the penis) contracted 3–4 times in 0.8 second intervals with younger adults. The distance semen traveled after expulsive release was 12–24 inches. Those in later adulthood also experienced 0.8 second intervals, but there were only 1–2 contractions per inteval. The ejaculatory distance was 3–5 inches.

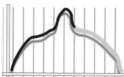

RESOLUTION

The return to the pre-arousal state took from minutes to hours in the younger population. It could be classified into two stages. In the older sample, there was only one stage. Pre-arousal states were achieved in seconds.

overgrowth can even press on the tubes that normally help void urine. As men get older, it can be increasingly difficult – and painful – to urinate. Although not necessarily a threat to health, the discomfort experienced is usually enough to require medical intervention.

Erections and sexual activity

One of the concerns often asked by men in later adulthood has to do with their sexual responsiveness. The subject of concern is the ability to produce and maintain erections – as well as effects on the sexual response cycle.

Males get erections because of nerves. Upon arousal, the neurons in a circuit called the parasympathetic nervous system divert more blood into the arteries of the penis. The increased input volume relative to the amount escaping expands the tissue and allows the erection to take place. Because there is great elasticity in penile tissue, the net effect is a temporary increase in the diameter and length.

As men get older, it is increasingly difficult to produce and then maintain the penis in an erect state. There are several reasons. There is increased growth of more good old inflexible connective tissue on the inner surfaces of the penis, generally beginning between the ages of 30 and 40. The veins and arteries in this region become more rigid, allowing less blood to flow through them. With less elastic tissue and less blood, the penis cannot maintain its erectile state for long. Even the normal erections occurring in sleep decrease after the age of 60.

So what?

As is true for females, later adulthood does not mean masculine cessation of sexual interest and activity. There are some changes obviously, but the same caveats given to the women must also be applied to men. Much of sex is psychological, and the anatomical alterations must be integrated with the person's own perceptions in order to arrive at the correct view.

There are physiological changes, as noted in several important studies. The four stages of sexual activity were compared between young men (20–40 years of age) with those in later adulthood (50–89). Here's what was found.

Excitement The time needed to obtain an erection after stimulation was 3–5 seconds in the younger sample and 10 seconds to several minutes in the older population.

Plateau Quick development of pressure for ejaculation was experienced in the younger population. The older population took longer to feel the pressure. Psychologically, those in later adulthood felt less of a need to ejaculate during a sexual episode or even across several episodes.

Orgasm In the younger men the penile urethra (the tube in the center of the penis) began contracting 3–4 times in 0.8-second intervals. The distance semen traveled after expulsive release was 12–24 inches. Those in later adulthood also experienced 0.8-second intervals, but there were only 1–2 contractions per interval. The ejaculatory distance was 3–5 inches.

Resolution The return to the pre-arousal state took from minutes to hours in the younger population. It could be classified into two stages. In the older sample, there was only one stage and it took only a couple of seconds to achieve it.

Conclusions

We cannot escape the fact that the sweeping biological changes that affect most of our body also affect our reproductive organs. We also cannot escape the fact that, in terms of enjoyment, such changes mean nothing.

Women and men age very differently here. Because sex hormones affect tissues throughout a woman's body, loss of these hormones affects more than just the reproductive organs. The aging of men's reproductive organs is much less pronounced. Rather than a sudden cessation, there is a gradual tapering off of reproductive viability. Like Giovanni Casanova, males can continue to sire children well into their sixth decade of life.

Whereas Casanova's body may have quit because of disease and medical malpractice, it would eventually have deteriorated and died even if he had lived like a saint. And we are once again left with questions like: what makes estradiol decide to reduce its levels in the body? Why do testes become unresponsive to FSH? Why does it take longer for both sexes to arrive at arousal and climax and resolution during sex?

The answer, of course, lies locked deep within the biology of individual cells. Hang on to those questions. In the next chapter of this book, we are going to find clues that regulate these cellular changes. And maybe even some that will help us uncover the key that winds up The Clock of Ages.

PART THREE

Why do we age?

INTRODUCTION

This last part of our tour opens up the back panels of The Clock of Ages and examines its ticking innards. To start this process, I would like to return to a discussion of the contents of human wills and testaments, a subject previously mentioned in the introduction to Part Two.

Wills reveal an interesting helplessnes in the human condition. We write them to exert authority over our possessions, because we are powerless to do anything about them once the inevitable occurs. Some of these wills make for some interesting reading, and not just as a repository for practical jokes. They can highlight some very human attitudes about the strange ambiguity of death.

For example, there was a will left by a wealthy banker who did not allow the following people any part of his vast estate: 'To my wife and her lover, I leave the knowledge I wasn't the fool she thought I was. To my son, I leave the pleasure of earning a living; for twenty-five years he thought the pleasure was mine.'

Not all wills are so unfriendly, of course. A business executive frustrated over his choice of professions left a generous donation with a very interesting request. 'All my life I wanted to be on the stage. Lack of talent prevented me from realizing that wish.' His entire estate was conditionally willed to a theater company. To get the money, the company had to make sure that the executive was decapitated, the tissue removed from his skull (which was subsequently bleached), and his skull cap used as the character Yorick in the play *Hamlet*. The theater did exactly as requested, and the money was theirs.

People have left all kinds of wills throughout history. The shortest on record belongs to a German, Karl Tausch. Dated January 19, 1967, the entire document reads: 'All to wife.'

The longest will in history was left by Mrs Frederica Cook, and is dated November 2, 1925. She actually had a relatively small estate, but her comments, best wishes, fond farewells and even some venom, took four bound volumes to deliver. It was roughly the size of War and Peace. It apparently was not read during the distribution of assets.

Though we might try somehow to outlive ourselves, whether we become a skull in a play or the author of an encyclopedia, we are only postponing the unquenchable ticks of The Clock of Ages. That death appears inevitable is testified by the fact that the oldest verifiable human on record lived 120 years, 237 days.

And that person is now dead too.

The factors that create our longevity are influenced by many things, including gender, race, socioeconomic status and even profession. Females live longer than males, Swedes live longer than Nigerians, rich people live longer than poor people, lawyers live longer than farmers. But the fact that longevity can be influenced at all begs an interesting question: how set in stone is our life span?

I say that because of an interesting series of facts we discussed in the last chapter. It begins with the obvious notion that aging occurs in organisms because some of their cells die. Or the cells change function and reduce their youth-giving contributions to the body. Skin cells start making melanin and we get age spots. Cells within the small intestine quit transporting calcium, and our bones pay for the loss. Certain cells in our joints stop making cartilage, or change the way it's manufactured, and this results in decreased mobility. Why does this occur? And if we could stop the cells from misbehaving, could we stop the damage?

In this last stop on our tour of The Clock of Ages, we will attempt to answer such questions. We will examine how individual molecules in our tissues work together to bring us the aging process. This means we are going to examine not the world of the organ, but the world of the cell. We will find that in order for the cell to survive the ravages of time, it will (a) need to sustain normal function by adapting or resisting environmental stresses, and (b) repair or replace damaged molecules and structures. Most fascinating of all, we will find that survival of some cells depends on their ability to stop the activation of a pre-programmed self-destruct mechanism.

As ever, caution must be exercised whenever we attempt to apply what goes on in a dish to what goes on in an organism. Experiments with cells in petri plates exist in a non-interactive environment. Nothing could be more diametrically opposed to the real world situation in our bodies, where tissues talk to each other with the frequency of teenagers on a telephone. The best use of cultured cells is to isolate a process that suggests a research direction in whole organisms. And then do research on whole organisms. Where possible such data will be described.

Looking at the mechanisms of cells has given us extraordinary insights into the overall process of aging. We have peered into the world of the gene, and used the powerful techniques of molecular biology to answer some tough questions. This research has some real spin-offs to other areas, such as cancer research, and the discovery of how embryos construct their tiny bodies from two cells. Most exciting of all, there is room to speculate on a wish concerning the business executive who donated his skull to the theater: the possibility of extending the human life span, not with last wills and testaments, but with real live genes.

12

A tale of two theories

We will begin our discussion of the root causes of aging by describing not the death of a cell, but the death of a king. The monarch was King Charles II of England (1686) and is a tragic case of deliberate if naive, error. We know about it because of the diary of Charles Scarburgh, the chief Physician to the King.

'I flocked quickly to the King's assistance,' Scarburgh wrote upon hearing of a sudden illness on the part of His Majesty. The king had been at his morning shave earlier in the day when he let out a terrific scream. He quickly collapsed into a quivering heap, rolling around the floor before slipping into unconsciousness (modern diagnosticians believe he suffered a sudden stroke that was accompanied by a seizure). Edmund King, a physician staying as a guest of the crown, was quickly summoned. He promptly adminstered emergency care, which consisted of cutting a slit in Charles' arm and withdrawing 16 ounces of blood. The call went out for Scarburgh, who took not only the best technology available to the 17th century, but also his diary.

The king had not responded to the emergency measures. Scarburgh, after consultation with six other professionals, decided that insufficient blood had been taken. The king's shoulder was cut in several places and an additional 8 ounces were extracted. Now the king moved slightly, rallying to a semi-conscious state, which convinced Scarburgh that the right treatment was being applied. If more fluids were extracted, he reasoned, the king might show even more improvement. A concoction of potassuim tartrate (now known to be a poison; even in those days it was used as a caustic in cloth coloring) was given to induce vomiting. This the king did with gusto. Pleased with his work, Scarburgh also delivered an enema.

Not surprisingly, the king slipped back into unconsciousness. He was given another enema and another batch of tartrate. This time the poor king did not respond. Scarburgh ordered Charles' head shaved and camphor (which can cause localized burns) and mustard plasters applied to the exposed scalp. The plasters contained cantharsis, a skin-absorbable irritant of the urinary tract. The king began to urinate uncontrollably. And perhaps not surprisingly, suddenly woke up.

The doctors were beside themselves with joy. He was given a powder to induce sneezing and another emetic to induce vomiting. As night was approaching, the doctors adminstered the most massive enema yet, to keep the bad humors flowing out consistently. In the morning the king was bled again, this time through the jugular veins. Another 10 ounces of blood was extracted.

Unfortunately, the king suffered more convulsions early the next day. He was bled again, and given a powder derived from the crushed skull of an 'innocent man.' The convulsions went away, but by the next day, he was exhausted and suffering from fluid loss. Not to be out-smarted, Scarburgh ordered another purgative, enema and bleeding. He even gave Charles quinine, in those days called the Jesuit's bark. The toxic amounts supplied to His Majesty made him grow worse. Scarburgh wrote: 'Alas, After an ill-fated night His Serene Majesty's strength seemed exhausted to such a degree that the whole assembly of physicians lost all hope and became despondent.'

Not to mention His Majesty.

Desperate now, the physicans threw at their sovereign everything 17th century medical creativity could muster. The king was bled nearly dry. He was given to eat a hellish concoction that contained 'extracts of all the herbs and animals of the kingdom.' When he could not swallow on his own, it was forced down his throat. He grew breathless and after the last bleeding, comatose. He died soon after, a monument to the resiliency and endurance of the human body. It had taken nearly a week of treatment to kill him.

By the standards of modern medicine, Scarburgh's medical program seems more like a prescription for the king's torture chambers than for the king's health. The protocol he and his colleagues followed, however, was in keeping with the best traditions of the time. The reason it seems so horrible was that it was so deliberate, even programmed. The fact that this sad state of 17th century medicine killed more people than it helped only makes this programming more tragic.

I would like to use the idea of errors and programmed death in another process with which King Charles II was acquainted, our present subject of the biology of human aging. This chapter provides background information for understanding aging at the molecular level. We will start by describing two theories of aging that incorporate the selective activation and deactivation of genes in cells. Then we will review some basic molecular biology, talking about how genes and proteins normally work in a healthy cell. A description of how genes 'know' when to turn on or off will conclude this chapter.

Two theories about aging

When one examines data about cellular aging, two large hypotheses are immediately confronted. In many ways, they are like Scarburgh's treatment of Charles II – a chronicle of errors unleashed in a deliberate program. I've named the two dominant theories Error Accumulation and Genetic Program Activation.

Error Accumulation

This theory rests on the assumption that as cells get older, they accumulate errors. The errors collect passively, due to the normal wear and tear of living within a sometimes hostile planet (scientists call these 'environmental insults'). Humans have normal repair mechanisms inside their cells, processes which can handle a certain amount of tissue damage. Aging is explained as an imperfect ability to repair all the breakdowns that normally occur. The large organism succumbs because of its inability to withstand relentless environmental insults.

Genetic Program Activation

This theory hypothesizes that the aging of organisms occurs because of a genetic conspiracy; that a much more deliberate breakdown occurs, with banks of 'suicide genes' turning on at specific times, making cells debilitate and tissues deteriorate. This theory is strengthened by the discovery of genetic sequences that can dramatically alter the life span of individual cells and even whole organisms. It also extracts ideas from developmental biology, discussed in an earlier chapter, in which programmed cell death is experienced normally as complex babies form out of simple sperms and eggs.

These two hypotheses are not mutually exclusive. As more genes and genetic processes are isolated, the boundaries separating indiscriminate error from deliberate program are blurring. In fact, as we'll see in future chapters, recent data suggest this bifurcated classification scheme works only if you don't look too closely. What both hypotheses have in common is a set of molecules whose behavior changes. As ever, changes in cellular function usually have something to do with genes and chromosomes. To understand what that means, we have to talk about the cellular environment.

A hard-cell setting

As you recall from the first chapter, you can think of a human cell a lot like you think of a fried egg. The yolk of the egg can be thought of as the nucleus of a cell; the white of the egg an area known as the cell's cytoplasm (Figure 37). The nucleus and the cytoplasm – this egg yolk and egg white – interact with each other in very important ways.

I once had a teacher describe to me the function of the nucleus: 'The nucleus,' she'd say, 'serves as a warehouse for all human genetic information.' It turns out that's not quite right. But the nucleus contains *most* of the information.

'This information can be divided into volumes,' she'd continue, 'just like an encyclopedia. There are 46 "volumes" of genetic information in a nucleus, which are called chromosomes. At a certain time point in the cell's life cycle, these chromsomes look like little Xs.' Sort of like alphabet soup that got stuck on one letter.

'The important point is that these chromosomes contain lots of information. When all 46 are placed end to end, there's almost 1.8 meters of genetic information available – and it all has to fit inside the nucleus. That's like taking 30 miles of your favorite fishing line and stuffing it into something the size of a blueberry. And it isn't just stuffed into the nucleus, like random cotton stuffed into a teddy bear. Rather it is folded into the nucleus, like a beautiful tablecloth might be folded into a drawer. This three-dimensional configuration is very important to the cell – it helps determine its career options – like whether it will become a nerve cell or a fat cell or what have you.'

THE ANATOMY OF THE
NUCLEUS
OF A HUMAN CELL

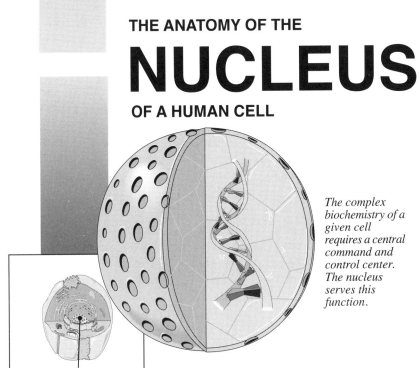

The complex biochemistry of a given cell requires a central command and control center. The nucleus serves this function.

NUCLEUS

FIGURE 37

The nucleus of a cell contains DNA, pictured above in its double-helical configuration. The outer structure of the nucleus is peppered with holes. These perforations are called nuclear pores and consist of groups of proteins. The pores act as border crossings, allowing the nucleus and the cytoplasm to transfer molecules back and forth.

The inside of the nucleus is also complex. It is composed of an inner framework, in the same way a human body has a skeleton. This framework is made of protein and is called the nucleoskeleton.

She went on to review something we discussed in Chapter 1. Chromosomes are made of the familiar molecule DNA, which looks like a flexible ladder that has been twisted at one end. She said that there are both active and inactive regions on human chromosomes. The active regions are called genes, something we termed 'traits' in an earlier time. Since chromosomes are made of DNA, a gene is simply an active region of DNA. The inactive regions are not called genes. 'We call it,' she said, looking at her watch, 'junk DNA.' She mumbled something about scientific arrogance, and then dismissed the class.

What did my teacher mean when she described the fact that there were active and inactive regions of DNA? It all has to do with the central purpose of genes. This purpose is not hard to understand. Genes possess encoded information. Cracking the code has been one of the greatest scientific achievements of all time (Figure 38).

And just what does that code say? Its most important communication, from our standpoint, can be summarized as follows. Genetic information says: *Make me a protein. A specific protein.* Different genes encode the information for different kinds of proteins. There are, on average, 3000 activatable genes on one chromosome. That means a given chromosome has the genetic information to create 3000 different proteins.

As we have seen from our discussion in the second chapter, proteins serve very important structural functions within the body. Elastin and collagen are proteins, for example. The cells that make up our muscles are proteins too.

But proteins also supervise most of our important bodily functions. We can hear music because of proteins. Molecules of food are shuttled from the small intestine into the blood stream because of proteins. Our eyes blink and our hearts beat and our lungs breathe because proteins are busy supervising *all* the relevant mechanisms. Without proteins, life on this planet as we know it – and indeed our own lives – would not exist. And without genes, there would be no proteins.

So how does a gene make a protein?

It took many experiments to answer the question about how code embedded within the DNA could supervise the manufacture of proteins. There seemed, at first, to be a contradiction between two known facts concerning the location of specific processes.

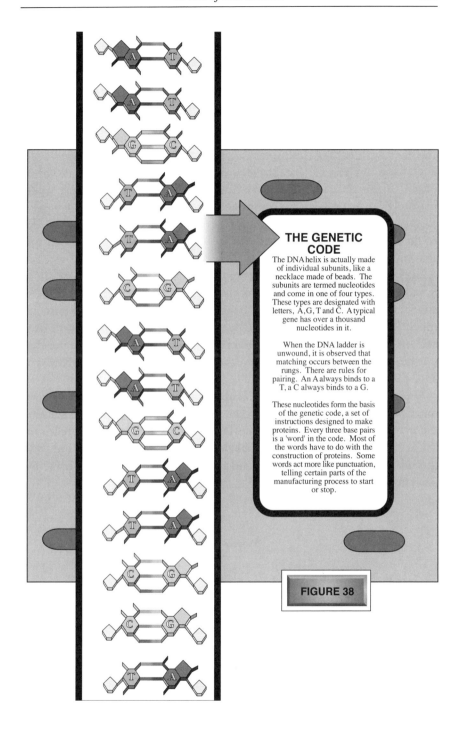

THE GENETIC CODE

The DNA helix is actually made of individual subunits, like a necklace made of beads. The subunits are termed nucleotides and come in one of four types. These types are designated with letters, A, G, T and C. A typical gene has over a thousand nucleotides in it.

When the DNA ladder is unwound, it is observed that matching occurs between the rungs. There are rules for pairing. An A always binds to a T, a C always binds to a G.

These nucleotides form the basis of the genetic code, a set of instructions designed to make proteins. Every three base pairs is a 'word' in the code. Most of the words have to do with the construction of proteins. Some words act more like punctuation, telling certain parts of the manufacturing process to start or stop.

FIGURE 38

(1) Genes are found only in the nucleus;
(2) Proteins are manufactured only in the cytoplasm.

There was an obvious contradiction. The genes couldn't move. They were anchored in their nuclei with no chance of escape. At the same time, no proteins could be made in the nucleus. That work was clearly done in the cytoplasm. The conclusion was that information was spilling out of the nucleus. The next question was: how did the genetic information get out of the nucleus so that proteins could be manufactured in the cytoplasm? If the genes couldn't leave their nuclear incarceration, they would be forced to write down their messages and somehow get them smuggled to the outside world. There would have to be a transport system that could deliver the information to the protein-manufacturing plants outside the nucleus.

Such logic was experimentally confirmed. To make a long story short, it was discovered that genes xerox their instructions onto molecular hard copy. The copy machine genes employ is actually a group of proteins called RNA polymerase. This polymerase has the job of binding to a gene and making a perfect, but quite mobile copy of what it sees. The message it creates is called, not surprisingly, messenger RNA. Scientists abbreviate it to mRNA (Figure 39).

Once the mRNA is made, it is handed over to the transportation department of the cell. The message is then spirited out of the nucleus and into the cytoplasm. Once there, the message finds its way to the protein synthesis factories of the cell. The molecular workers within the factory read the instructions on the message and, to exacting specifications, make the protein the message describes.

Activation

By asking for a gene's ability to make mRNA, we are really asking 'is it active?' The reason we pose this question is that a lot of genes are inactive in a given cell. There are genes that are never turned off in a given cell and there are genes that are never turned on. There are even genes that are active only for a short while and then become turned off completely.

To understand the relevance of this to our discussion of the aging process, let us consider an interaction that always occurs between parents and children. This particular event happened to a friend of mine.

The time when young children ask about the facts of life is a moment many parents dread. In my friend's family, the conversation came about because of the father's laughter at a newspaper article. The headlines were:

Lonely Deer Finds Company with Concrete Yard Animals.

The article described the amorous interactions of a flesh and blood young male deer with two life-sized concrete lawn ornaments, fashioned in the shape of a male and female deer. At first the confused young buck charged the male, but quickly gave up the battle in the absence of reciprocating hostility. Sensing permission, the young buck then approached the concrete female, making love to it for several hours. At one point, the randy ruminant even mounted the male. A car's light eventually chased the deer away. After reading this story, my friend put his paper down and started to laugh uncontrolably.

'Why is that so funny, Daddy?' his little girl questioned him as he related the story to his wife. 'Was he trying to make babies?' What followed was a conversation about reproduction, how animals have to join in order to create babies, just like humans do. Her father described in patient terms how little babies grow up to become adults, who in turn have babies. The father returned to his paper, happy at satisfying his daughter's curiousity. But then she asked a bombshell of a question.

'Why do babies look like babies and daddies look like daddies, Daddy?' The father began an answer, and ended with a blank look. He wasn't a biologist and hadn't the slightest notion about the complex interaction of genes with anatomical variety – and how everybody gets older. He gave a sheepish shrug to his daughter and promptly gave me a phone call.

My friend and I talked for hours.

Developing answers

The little girl's question aims right at the heart of a tremendously exciting inquiry in modern biology. What makes cells differentiate so that we stop looking like babies and start looking like adults? Why do some cells become cheek cells and stay that way, others nerves, others blood vessels? Why do these changes affect our appearance, from our featherweight days as babies to our overweight days as middle-aged adults?

A genetic problem solved

Genes, which encode information to make proteins, can't leave the nucleus. The protein manufacturing centers, which exist in the cytoplasm, can't get into the nucleus. How are proteins made? Illustrated below is the cellular solution, the creation of a mobile messenger.

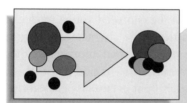

1 RNA polymerase assembles into a single complex from a group of smaller subunits.

Promotor
Beginning of gene

2 The complex searches for a promoter region just upstream from the target gene. Once it finds the promotor, the RNA polymerase binds to it.

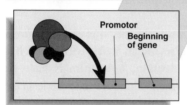

New mRNA chain

3 The RNA polymerase 'reads' one strand of DNA. It creates a mobile messenger, called mRNA, from what it sees.

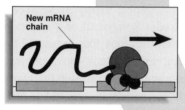

4 After extensive editing, the mRNA is ready to leave the nucleus. With the help of escort molecules, the mRNA exits via the nuclear pore. It then finds the...

FIGURE 39

...PROTEIN MANUFACTURING PLANT.

The protein manufacturing plant inside cells is called the ribosome. A complex mixture of molecules, the ribosome receives the mRNA from the nucleus. It 'reads' the instructions embedded within the message, creating a protein according to the instructions. The ribosome doesn't look like a heavy industrial plant of course. Pictured to the right is a cartoon approximation of its overall shape.

This question is fascinating because of an obvious secret, not surprisingly, having to do with genes. If you were to examine the nucleus inside one of the cells of your cheek, you would find all the genetic information necessary to establish and maintain a cheek cell. It would not be a nerve cell, nor would it become a cell that makes bone. The information makes sure that the cell does everything a cheek cell is supposed to do. This seems intuitively obvious, especially considering the command and control functions of the genes within the nucleus.

But if you kept looking at the genes inside that cheek cell, you would find an astonishing thing. You would find not only the genetic information necessary to make your cheek, you would also find the genes that code for the proteins of your heart. And liver. And pancreas. Your cheek cell contains genes that help in the construction of a human skull, knee joints and kidneys. This innocent little bag of molecules in your mouth contains the complete instructions to make all the proteins and perform *all* the functions of your body.

It gets worse. *Every cell of your body contains the entire blueprint for the protein manufacture of a new you.* Since you have 60 trillion cells, there are also 60 trillion complete blueprints. All 46 chromosomes are present, all 150 000 are available for work in every cell.

If this is true (and it is), you can ask a question related to my friend's little girl. Why is a cheek cell always a cheek cell? Why isn't it beating like a heart? Or thinking like a nerve? All the genes are there, all the information necessary to make some pretty exotic tissue is present. What makes a cheek cell so loyal to its job that even if it's wounded, the inside of the mouth won't grow a lung or a liver, but merely more cheek cells?

Selective activation

The answer is beginning to be understood – and it is the stuff of science fiction. It turns out that all the genes necessary to make a cheek cell are active in a cheek cell. By active, I simply mean that those xeroxing RNA polymerases are able to find the gene, make a messenger from it and send the codes off to the protein synthesis factory. All the other genes – those which can make brains and pancreases and intestines – are silent. Turned off. That's the reason why a cheek cell is always a cheek cell. Instructions have been burned into its chemistry never to turn on the irrelevant genes.

This same pattern of activation/inactivation is seen in cells throughout the body. The genes necessary to keep a liver cell functioning within the body are active in a liver cell, but the co-resident nerve genes are silent. Pancreatic cells have activated pancreatic genes; everything else is quiet. The specialized functions of all our tissues occur because the genes are selectively recruited. All the other genes are told to keep quiet. The secret to finding the functions of a given cell boils down to asking which genes are making mRNA and which ones are not.

And the secret to its control is: who makes the decisions to turn them on and off?

There are many ways genes become activated. To understand the control decisions, we will examine three overall mechanisms cells employ to wake up genes (Figure 40).

And we will begin the entire discussion by talking about airplanes.

Crowded runways

I've often been told by pilot friends of mine that the two most hazardous times to be aboard an aircraft are during take-offs and landings. As if the physical requirements needed to get large airplanes in and out of crowded airports aren't enough of a problem, one sometimes has to worry about biological problems as well.

Consider the case of a DC-10, taking off from a runway at New York's Kennedy International. As it was charging down the runway, its throttle wide open, a flock of seagulls suddenly appeared out of nowhere. The panicked pilot immediately slowed the plane down and slammed on the brakes. The stress applied to the airplane was too much to take, however, and an engine on the right wing fell to the ground. The plane immediately caught on fire. Once the plane had come to a full stop, all 139 passengers and crew were evacuated. And just in time. The plane exploded in a massive fireball soon after the passengers returned to the terminal.

These biological hazards occur during landing as well as take-off. During a routine military flight, Lt Col Sam Carter had been given permission to land at Jacksonville International Airport in Florida. As is true with all airports, a homing beacon was sent from the control tower to the airplane. A device on the airplane was activated that told the colonel the location of the runway.

Flying a $16 million F-18 fighter, the pilot was trained to land under the most adverse conditions a battlefield might offer. Unfortunately, no

How genes are turned on

Turning on a gene means getting it copied into mRNA. Here are three ways cells make their genes active.

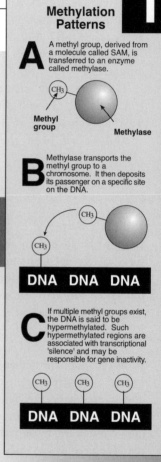

Transcription defined

There are many genes in a typical human cell. Only some of them are open and available for work. The open genes that coax the cell's xerox machinery, RNA polymerase, to copy them are termed 'active.' All the other genes are inactive. This copying function is called transcription. RNA polymerase is a mixed complex of proteins whose job is to find open genes and transcribe them.

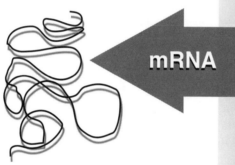

Why the selectivity?

This preferential stimulation confers on cells unique properties. Why is that necessary? Nearly every cell in the human body contains a complete set of human genes. The genes in the skin contain the information necessary to make the proteins in your brains, the cells of your liver contain 'heart genes' and so on. Selective activation ensures that specialization occurs. Genes needed to make a stomach are active in stomach cells and not in muscle tissue, for example. Understanding what activates and deactivates genes in cells sheds great insight into how human beings are genetically constructed.

1

Methylation Patterns

A A methyl group, derived from a molecule called SAM, is transferred to an enzyme called methylase.

Methyl group

Methylase

B Methylase transports the methyl group to a chromosome. It then deposits its passenger on a specific site on the DNA.

CH3

CH3

DNA   DNA   DNA

C If multiple methyl groups exist, the DNA is said to be hypermethylated. Such hypermethylated regions are associated with transcriptional 'silence' and may be responsible for gene inactivity.

CH3   CH3   CH3

DNA   DNA   DNA

FIGURE 40

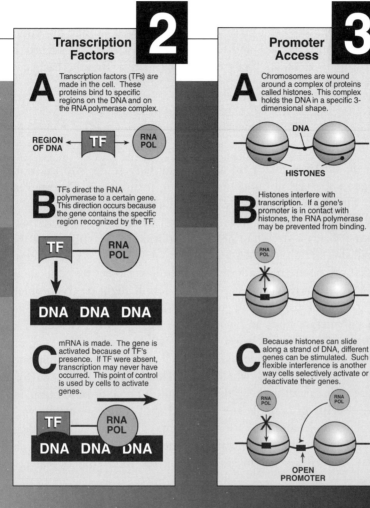

Transcription Factors 2

A Transcription factors (TFs) are made in the cell. These proteins bind to specific regions on the DNA and on the RNA polymerase complex.

REGION OF DNA → TF → RNA POL

B TFs direct the RNA polymerase to a certain gene. This direction occurs because the gene contains the specific region recognized by the TF.

TF   RNA POL

DNA   DNA   DNA

C mRNA is made. The gene is activated because of TF's presence. If TF were absent, transcription may never have occurred. This point of control is used by cells to activate genes.

TF   RNA POL

DNA   DNA   DNA

Promoter Access 3

A Chromosomes are wound around a complex of proteins called histones. This complex holds the DNA in a specific 3-dimensional shape.

DNA

HISTONES

B Histones interfere with transcription. If a gene's promoter is in contact with histones, the RNA polymerase may be prevented from binding.

RNA POL

C Because histones can slide along a strand of DNA, different genes can be stimulated. Such flexible interference is another way cells selectively activate or deactivate their genes.

RNA POL   RNA POL

OPEN PROMOTER

amount of training could prepare Carter for the greeting he was about to receive. As his plane touched down on the runway, a pair of wild pigs suddenly appeared from nowhere. They walked directly in front of the landing aircraft. A frightful collision occurred, one which damaged the jet and slaughtered the pigs. The pilot managed to take his injured airplane back into the sky, and then he ejected. Carter was slightly bruised and the pigs were, in the words of his jocular friends, barbecued and ready to eat.

Promotors

What do airplanes taking off and landing have to do with understanding how genes awaken in human cells? A lot, it turns out. The relevance has to do with how RNA polymerase finds a copy-ready gene.

Each human gene has what might best be termed a landing field. This runway is really a stretch of DNA that physically lies in front of the gene. We call this stretch of DNA a promoter. When an RNA polymerase finds an open promoter (near a gene) the complex can act just like an airplane that has been given permission to land. It will alight on the promoter and then go to work copying the gene. That's one way to activate a gene. The RNA polymerase finds an open gene by asking: 'Am I cleared to land on the promoter?' If it is, then a message will be made.

This promoter sequence allows the cell an opportunity to control which genes get activated and which ones remain closed. The control exists because cells can either keep their promoters open, or send in the molecular equivalents of seagulls and pigs and shut the promoter down. These controlling molecules are called repressors. They can actually stick to the promoter, rendering it inoperative in the same way animals can disable a runway. The RNA polymerase, sensing the interference, will refuse to land.

Sometimes activating a gene is more complicated than just finding an open promoter. In many instances, the RNA polymerase can't land without a homing beacon near the promoter region. After all, the nucleus is a big place and there are thousands of genes. Just like an airplane, the polymerase sometimes needs to be outfitted with a device that can recognize the beacon near the genetic runway.

The polymerase is outfitted with just such a mechanism. Called transcription factors, they are proteins that actually bind to the polymerase and guide it to the proper runway on the gene. How do they know which runway to use? They look for homing beacons, just like airplanes do. Only these are embedded in a chromosome rather than in a control tower. Once the polymerase is guided to the right spot, it lands on the promoter and activates the gene.

Popcorn-strung Christmas ornaments

This airport method of finding the right promoter is only one level of control. It is a powerful, quite flexible way of orchestrating genetic programs. Without clearing away a promoter, the RNA polymerase won't land on a gene. But even that is no guarantee. Sometimes the proper transcription factors are needed before the complex receives permission to land.

If the seagulls and the pigs aren't cleared off the runway, no landing takes place. But that clearing is no easy step. Many promoters, in the immensely crowded universe of the nucleus, are covered up. To understand this, as well as the next level of activation, consider for a moment popcorn ornaments and Christmas time.

When I was little, our family used to decorate the tree with popcorn strings. Mom and Dad always had these beautiful multi-colored threads. The ideas was to attach the popcorn to the thread – and then hang the chain on the tree. We would pop the popcorn, eat as much as possible and then start making the decorations. There were many ways I tried to get the popcorn on the thread, and my parents, sensing the experimenter, would just let me try and fail.

And I tried everything. I got out some glue and stuck the popcorn to the thread. It always fell off. I tried tape, which made my ornament look really messy – and the popcorn still fell off. I even tried wrapping the thread several times around the kernel. That was always the prettiest thing to do, because I could see the various colors of the thread around the white popcorn. And I discovered if I wrapped the thread around it only a couple of times, I could actually 'slide' the kernel along the thread's length. This would change the colors visible on the face of the popcorn. I would get lost trying to make a whole string of slideable popcorn decorations.

As you might imagine, I was not put in charge of making the popcorn ornaments for our Christmas tree.

The role of histones

What do popcorn ornaments have to do with gene activation? I would like to use the analogy to describe the three-dimensional structure of the DNA as it is coiled in the nucleus of the cell. This internal configuration has a lot to do with gene activation, and with popcorn tree decorations.

When we think of chromosomes, we generally imagine those little Xs lined up like tiny soldiers in the middle of the nucleus. But that is only how they look for a very short period of time – when the cell is getting ready to divide. The rest of the time they are coiled inside the nucleus like a curled ball of thread. The nucleus housing 46 of these balls is a very jumbled looking place.

Don't let its appearance fool you, however. The curled ball structure is highly organized. If one looks at the chromosomes carefully, one sees that it is not just made up of molecular string. It actually has regularly spaced 'lumps' on it, the same way a thread that is part of a Christmas popcorn decoration does. These lumps are made of a complex of proteins we have collectively termed histones. The DNA is wound around these lumps in the same way that I tried winding thread around the popcorn. All human DNA has histones associated with it, spaced at regular intervals, controlling the three-dimensional structure of the chromosome.

What do the presence of histones have to do with gene activation? Its relevance is related to the way the DNA is coiled around the lumps. The DNA helix wraps around the histones exactly twice. This means that, like my popcorn decoration, the histone is 'slideable.' A given histone has the ability to slide over different regions of DNA. The lumps can cover one area for a while, and then, if moved, will expose that area and cover a new one.

Why am I using words like cover and expose with DNA? You recall that in order for a gene to be activated, the promoter (that runway we discussed earlier) must be 'open.' It has to be cleared of molecular seagulls and pigs. Those obstacles can be many things, one of them being histones. Here's the rule: If a histone is in the way, the gene is inactive.

You don't have to think too hard to see that this is another way of controlling gene expression. Just block the promoter and no gene will be activated. Unblock the promoter and the gene it regulates becomes susceptible. Cells do this a lot. That's why the three-dimensional configuration of DNA is so vital to the career options of the cell.

Stopping a prickly gene

I would like to explain one more way in which genes can be activated. This method reminds me of the great power of very tiny molecules over large organisms. We have discovered that when promoters are not obstructed, the RNA polymerase can bind to the DNA and make a message. The idea in that system is to make something active by first making it available, by first making it 'open.' There is a mechanism cells employ which can render even the most open gene inactive, however. The mechanism is called methylation. Its role in gene regulation is discussed below.

In the chemical world, there is a molecule called a methyl group. Physically, it looks like a carbon porcupine, with several hydrogens stuck to the carbon atom like quills. And they work just like them, too. When normal porcupine quills brush up against something, the quills stick. When large biological molecules brush up against methyl groups, the groups can stick to them, too. But it isn't just a nasty practical joke the cell likes to play. In the odd world of cellular regulation, transfer of methyl groups to a complex molecule is an important point of control.

You might be asking what this molecular pricking has to do with DNA. The answer is 'plenty.' There are molecules that will take methyl groups and stick them onto genes. This doesn't 'hurt' the DNA molecules; the gene will just take the methyl group and wear it like a red badge of courage. The presence of the methyl group is not neutral, however. Silent regions of the chromosome are usually associated with lots of them. If a gene has been stuck with methyl groups, it is often rendered inactive. No mRNA will be made.

To be fair, the exact role of methyl groups and gene activity isn't clearcut. The presence of these little darts and the silence of the genes to which they stick is primarily associative. But their appearance on DNA, especially when compared to that of the other control points mentioned previously, demonstrates the creative number of ways cells selectively activate and deactivate their genes.

Conclusions

Taken together, we see that the world of molecular biology is a complex, fascinating interaction between many tiny players. The lilliputian genes, permanently chained to their nuclear dungeon, must somehow communicate their protein-making information to the cytoplasm. They do this by employing a molecular xerox machine, the now familiar RNA polymerase, to copy the instructions. This copy, this mRNA, can leave the nucleus and communicate with the rest of the world.

This process of mRNA manufacture, as we have seen, is under tight control. The reason is that there is so much genetic redundancy, with every cell carrying the blueprints of the entire organism. The solution to keeping us human is to create many levels of genetic regulation which in turn produces different job descriptions in our cells. Some control occurs by making accessible or rendering inoperable those promoters we discussed. This can be done by changing the three-dimensional configuration of the DNA, or by requiring the polymerase to carry certain homing beacon proteins, those aforementioned transcription factors. The raw DNA itself may be controlled by simply sticking it with the quill-like methyl groups.

What does all this have to do with the aging process? As mentioned earlier, growing old ultimately comes down to asking why cells decide to quit working. Two theories have been put forward to answer the question, named earlier as the Error Accumuluation and Genetic Program Activation theories. This chapter has given us the background necessary to discuss these ideas in terms of genes turning on and off. We must ask whether aging in cells is the inadvertent deactivation of some important genes, the sudden activation of a deliberate death program, or perhaps both.

These questions, couched differently for another reason in another century, probably ran through Charles II's head in the week it took him to die.

13

Error accumulation

Oscar Wilde died primarily because of a jail term.

This ignominious demise seemed an ironic contrast to his career as a powerful, clever writer, one who achieved success fairly early in life. His lightning-like wit thundered across the great skies of literary London. He became their instant darling, writing poems and plays deliriously received by large, enthusiastic audiences. Wilde was the quintessential 'dandy,' wearing outrageous ties, lacy frocks and multi-colored coats that glittered in the stage lights.

Oscar Wilde's downfall occurred primarily because of his sexuality. By the time he was 32, he was married, parental and bored. He began a sexual relationship with a teenaged boy named Robert Ross. This was the inaugural liaison of a pattern which culminated in a dalliance with Lord Alfred Douglas, the son of a very prominent English family. Douglas's father, the Marquess of Queensberry, discovered the relationship and wrote Wilde a note about it. In the mistake that would cost the writer his life, Oscar Wilde threw a lawsuit at the father. The marquess was used to fighting (this was the same gentleman who defined the Queensberry rules for boxing), and he countercharged that Wilde had numerous liaisons with young boys. The playwright was arrested and, in a scandal that rocked the literary world of the time, put on trial.

In court, things at first went well for the playwright. His wit sparkled with characteristic brilliance even upon cross-examination. The packed courtroom became as much a staged play as a criminal trial, with Wilde delivering the lines to a very delighted London intelligentsia. And yet, though the writer recognized the importance of being earnest, the evidence against him was overwhelming. Street boys testified to his excesses. Incriminating love letters from the writer's own hand to various men were produced as evidence. It was too much even for *his* wit to overcome. The jury deliberated for less than three hours before returning the verdict of guilty. To the shocked faces and proper countenances of Victorian England, Wilde was sentenced to two years in jail.

It might as well have been a death sentence.

His reputation and career were now in shambles, and he would never recover from the shame. His two years in prison were only a mock foreshadowing of the pain he would receive when he was set free. One day, while reading a book in jail, Wilde inadvertently slipped on a stone floor. He fell head first and damaged his ear. The ensuing infections contributed greatly to his demise.

Wilde lived two more years after he was released from prison. They were not happy ones. The writer, professionally annihilated and financially broke, became something of a wandering beggar. He began to drink heavily. He wandered around the European continent, cadging drinks from old friends, enduring the disdain and disgust of a public who, only a couple of years before, had hung on his every word.

The writer's ear infection grew worse, and in October 1900, he underwent surgery. While the exact nature of the operation is unknown, Oscar Wilde's health did not improve. There is speculation that the treatment may have caused his infection to spread from his ear into the lining of the brain, a tissue called the meninges (an infection of the meninges is called meningitis). There is some evidence that he had much earlier contracted syphilis as well, which may have hastened his decline. Regardless of the origin of the diseases he suffered, Oscar Wilde began failing fast.

By late November, the writer was delirious, conscious one moment and unconscious the next. The room he rented had garish wallpaper and Wilde is said to have murmered 'This wallpaper is killing me. One of us has got to go.'

It would not be the decoration.

This chapter

Oscar Wilde experienced the painful death of tissue degeneration. Unknown to the medical world of the time, lifestyle and health have to be generously treated in order to keep an immune system in fighting shape. The writer experienced neither, and his body slowly ground to a halt. It did not help that he was traumatized in later adulthood, at a time when other natural forces are already applying stress to physiological processes.

The idea of accumulated biological degeneration lies at the heart of this chapter. As we will discuss in a moment, there are researchers who believe aging occurs through a collection of biological errors. As marvellous as human biology is, the structural proteins, enzymes and tissues don't always behave as they're supposed to. Like the people that possess them, molecules sometimes make mistakes in the jobs they perform. These errors can have deleterious consequences for the organism. Some scientists believe that one of those consequences is the aging process.

In this chapter we are going to explore observations that support this idea of inadvertent degeneration. We will start by describing a reactive set of molecules termed free radicals. Then we will discuss genes deeply involved in cellular reactions to stress, the so-called heat-shock genes. The role of other molecules in the aging process, ranging from sugars to proteins to DNA, will be explained in terms of that Error Accumulation theory we discussed in the last chapter. We will end by describing the role of this theory in current biological research.

By examining these many processes, we will encounter the awesome power of biological excess. The overabundance of waste products can affect normal processes that, if left alone, would *never* age. A surplus of genetic mistakes can cripple tissues and result in the same tragic consequences. Since everything in our bodies is related, a profusion of errors in only a small number of cells can turn our world topsy turvy. 'Nothing succeeds like excess,' penned Oscar Wilde, who at the time, was describing his lifestyle at the pinnacle of his career. This author was as prophetic about the agonies of death as he was descriptive about the pleasures of life.

The power of electrons

In this section, we are going to take our first journey into the world of molecular aging. In previous chapters, we discussed natural changes in large collections of tissues, especially organs. In the last chapter, we described in basic terms the mechanisms of gene activation in a cellular world. Now we are going to describe the processes of aging inside a single cell. To start off our discussion, we need to know something about oxygen. And to understand what that has to do with aging, I would like to describe a little boy's experiment.

Jim, a childhood friend of mine, once learned an amazing scientific lesson. This learning experience involved the fact that both sugar and salt could magically disappear into a freshly brewed cup of coffee. The lesson took place on a summer's day in the 60s, the kind of day when parents took their kids to the neighbor's for some refreshing gossip.

Jim's mother participated in these migratory rites with some gusto, and Jim hated it. He was bored with the topics, or just plainly did not understand them, especially when the parents would speak in low whispers. On one of these trips, little Jim decided to relieve his boredom by performing a scientific experiment. While Mom and her friend went into another room to view some photographs, Jim found the salt shaker and the sugar bowl. He quietly poured the sugar out of the bowl and replaced it with the salt. And then he waited.

When the women returned, they poured the coffee into their cups. Both mothers liked their coffee with cream and sugar; absentmindedly, they doctored up their beverages. With the first sip, they noticed that something was terribly wrong. They also noticed Jim had started laughing. Jim's mother looked at her friend and they both nodded, not with shock, but with deliberation. In a concerted action Jim later described as 'too fast,' the friend grabbed the little boy and Mom proceeded to induce rather pronounced edema on Jim's bottom. She never took him to see her friend again. And he never forgot this scientific lesson, knowledge that would serve him well as a fraternity brother in later years.

Radical formation

The ability of coffee to hide its alteration with salt brings up a lesson for us too. Molecular structures can 'dissolve' in fluids, like the salt in the mother's coffee. Believe it or not, gases like oxygen can dissolve in fluids too. The very reason we can utilize oxygen at all is because it can dissolve in the water of our bodies. That way, it can be transported by our blood to the various tissues and perform its duties.

But what duties? Why do we need dissolved oxygen, floating around our blood stream like salt molecules in coffee? The answer has to do with waste disposal. You may recall from earlier chapters the role of mitochondria in human cells. These organelles act like tiny batteries, producing energy the cell utilizes for its various chemical reactions. They even have their own sets of genetic information, a circular chromosome containing important genes.

Like most manufacturing processes, this energy-production factory also produces toxic waste. This waste isn't smoke from a chimney or some kind of odd organic molecule, however. The waste is simply excess electrons, those cloud-like charges that usually orbit around an atom. Something has to be done about the excess electrons, because if they hang around, they can cause a lot of trouble to a cell.

The cell indeed has found a way to clean up the electron mess. It employs molecular oxygen, a gas that consists of two oxygen molecules held together in atomic matrimony. We symbolize this happy union by writing O_2. The clean-up occurs when a pair of waste-product electrons crash into the couple and separate them. Combined with other molecules (specifically two excess hydrogen ions, which are also lying around the cell), this unstable intermediate will quickly form water. The toxic electrons are thus absorbed into the water molecule, which can either be harmlessly excreted or used for other purposes. This transformation of hazardous waste electrons into docile water is a critical reaction. And oxygen is intimately involved. If the process ever broke down, we would be in serious trouble.

The problem, of course, is that sometimes it does break down.

Occasionally, those electrons do not crash into an oxygen pair and make water. Instead, the electrons crash into other atoms. This collision creates specific molecules, each carrying a wayward electron ('unpaired electron' in the jargon). Molecules with unpaired electrons are called free radicals (Figure 41). These free radicals are highly reactive and can do tremendous damage to the cell once they are formed. Because there are actually several types of oxygen-laden free radicals, scientists have been collectively calling them reactive oxygen species, or ROS molecules. We will too.

Tiny toxic dumps

The problems created by storing toxic ROS molecules are the same as encountered when storing any kind of hazardous waste. If ROS molecules are allowed to hang around inside a cell, they can create a tremendous amount of damage. Like a lot of industrial waste, these reactive species can indiscriminately damage anything that gets in their way – proteins, fats, RNA – even DNA (Figure 42). The effect is cumulative and a serious problem will occur if ROS numbers get too high. And they can get too

FREE RADICAL FORMATION

Free radicals are a form of toxic waste associated with the aging process. Here's what they are and how they form.

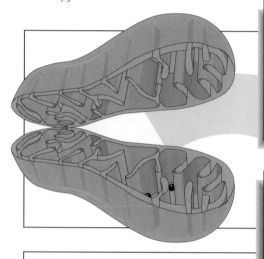

1 Free radicals are formed in the cell's mitochondria. This bean shaped organelle, split cross-ways in the illustration to the left, functions like a tiny battery. ATP is the chemical that stores the energy.

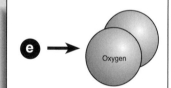
electrons

2 This power generation, like most manufacturing processes, produces toxic waste. The waste is in the form of excess electrons.

WHEN IT BREAKS DOWN...

The oxygen absorbs the offending electrons in the following way: it reacts with excess hydrogen in the cell. The combination of oxygen, hydrogen and electrons form a water molecule. This water is subsequently excreted or used for other purposes.

Sometimes this process breaks down and the electron binds to other molecules in the cell. Such chemicals possessing unpaired electrons are called free radicals. Free radicals are highly reactive and cause a lot of internal cellular damage. Oxygenated free radicals are also called ROS (Reactive Oxygen Species) molecules.

3 The reason we breathe air is to get rid of this toxic waste. Oxygen, like a sponge, normally soaks up excess electrons.

FIGURE 41

RADICAL DAMAGE

ROS (reactive oxygen species) molecules can do a lot of damage to the interior of cells – and cells have evolved mechanisms to fight them. Here's an outline of the battle.

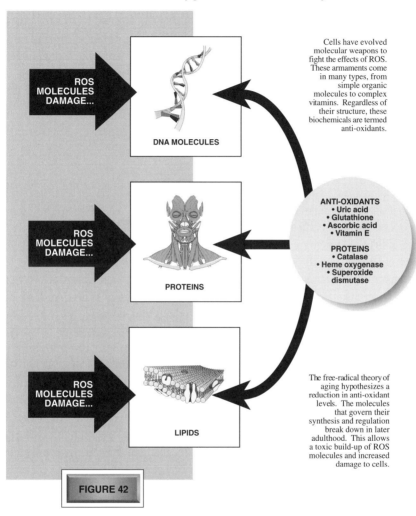

ROS MOLECULES DAMAGE...

DNA MOLECULES

ROS MOLECULES DAMAGE...

PROTEINS

ROS MOLECULES DAMAGE...

LIPIDS

Cells have evolved molecular weapons to fight the effects of ROS. These armaments come in many types, from simple organic molecules to complex vitamins. Regardless of their structure, these biochemicals are termed anti-oxidants.

ANTI-OXIDANTS
• Uric acid
• Glutathione
• Ascorbic acid
• Vitamin E

PROTEINS
• Catalase
• Heme oxygenase
• Superoxide dismutase

The free-radical theory of aging hypothesizes a reduction in anti-oxidant levels. The molecules that govern their synthesis and regulation break down in later adulthood. This allows a toxic build-up of ROS molecules and increased damage to cells.

FIGURE 42

high. It's been estimated that 1–4% of all the oxygen gobbled up by the mitochondria is converted into ROS molecules.

Not surprisingly, the cell has responded to this threat by creating various molecular 'firefighters.' Or perhaps it is better to say 'assassins.' The job of these heroic – if violent – biochemicals is to destroy ROS molecules. Some of these agents are chemical, ranging from ominous-sounding substances like uric acid to the friendly and familiar vitamin E. Collectively, molecules that destroy free radicals are called anti-oxidants. A number of proteins, created by real live genes, also munch on free radicals. They have exotic names like catalase and superoxide dismutase.

Relevance to the section

What does the formation of ROS molecules have to do with the aging process? And specifically, the Error Hypothesis? As we have discussed, the ability to destroy the functions of individual cells is one way to debilitate entire tissues. During the aging process, we have described that cells in specific tissues begin to die, or at least change their function. When that happens, they lose the ability to perform certain tasks. And so do we. The overall question has been: why does that occur?

To understand how ROS molecules can affect these tissues, we have to go back to that previously mentioned cellular organelle, the mitochondrion. One way to destroy or dramatically alter the function of a living cell is to take away its energy source. Annihilate most of a cell's mitochondria and the cell will be crippled. How can a mitochondrion be killed? Simply by taking shots at its supervising chromosome, that circular bit of DNA we discussed earlier. If the genetic information in the organelle is rendered useless, the cell will not have the energy to produce the various molecules needed to perform its special job.

In aging, that's exactly what occurs. The incidence of mutations and entire deletions in mitochondrial DNA increases as cells age. Something is clobbering the genetic information and as a result, less energy is produced. This has disastrous effects. If you're a skin cell, it may mean you can't produce the collagen and elastin necessary to keep the epidermis young and smooth. If you're a cell in the small intestine, it may mean you can't absorb all the nutrients you did in a younger year.

The culprit

What does this kind of damage have to do with ROS molecules? The breakdown in cellular safety nets probably accounts for the damage mitochondrial DNA suffers as we age. The likely culprits for this cruel destruction are ROS molecules, first generated in the mitochondria. There is evidence that these free radicals accumulate in these little batteries (and in other places, too). It is known that ROS molecules have a propensity to eat mitochondrial DNA for lunch. If they hang around for long periods of time, their harmful presence may destroy some gene sequences and mutate others.

Whereas the above mechanism is somewhat associative (and therefore only 'suggested'), there is direct evidence that such build-up affects aging. Recently, a group of genetic engineers created a special group of fruit flies. These flies were filled with those 'firefighter' genes we talked about earlier, those genes that can destroy ROS molecules on sight. The question they asked was this: if a fruit fly had an abundance of ROS-annihilating firefighters, how would their aging be affected? Fruit flies produce a lot of ROS molecules, and they don't normally have such an abundance of help. The researchers transferred these ROS-munching genes into the fruit flies, and then asked 'how long will they live?'

The answer was astonishing. The fruit flies lived more than a third longer than they should have. Their age-related loss of physical abilities was delayed substantially. They had fewer damaged proteins, molecules normally hit hard by ROS.

This result provided direct support for the involvement of free radicals in aging. And produced some interesting insights for ourselves. If we can apply fruit-fly data to human biology, it would seem that Error Accumulation is a big deal. Perhaps our tissues suffer damage with age, because there is a build-up of ROS molecules in our bodies. As the level of ROS grows, we suffer increased damage to our tissues. It is known, for example, that ROS-injured proteins accumulate with age in certain tissues, notably our intestines, liver and kidneys. But not just proteins. Fats, RNA, DNA – everything gets damaged.

If these data turn out to be relevant to more than just insects, there are lots of questions for the future: Why do ROS molecules *accumulate* with age and what can we do about it? One obvious conjecture is that the firefighter genes, which normally control the levels of ROS, get switched off for some reason. The result? ROS levels rise with age, because we just don't have enough firefighters to fight them off. If so, a 'treatment' for old age might be considered. What if ROS-firefighters were added to humans? Would that halt part of the aging process?

While that may be an interesting idea, for now it must remain only a thought problem. Nothing in biology has simple solutions, even considering ROS levels and fruit flies. Aging, after all, is a multi-factorial process. But the fact that ROS molecules accumulate only in later adulthood – when we're younger, this just doesn't happen – means that a regulatory process has begun to break down. Have we started to isolate that process? Do we know the life span of some of the firefighters? We are beginning to answer such questions about the regulation of Error Accumulation, and in the next section, we will explore them.

A heat-shocked lawyer

In order to talk about one particular ROS-rescuing protein, I would like to divert our attention away from the intracellular environment and talk about a living organism. A lawyer, to be specific, a controversial criminal attorney named N. Graves Thomas.

Mr Thomas lived in Shreveport, Louisiana, and was very well known. He came from a family of criminal justice professionals, including several judges and lawyers. He made a career of defending often notorious clients, including mob figures, drug pushers, thieves who killed policemen – even other lawyers. He became known for his strong loyalties to clients and powerful courtroom techniques. It wasn't a positive reputation, however. His peers described him as a firebrand and a maverick.

One day, Thomas, a young woman and three other friends were out water skiing on Lake Bisteneau. Heavy thunderstorms had buffeted much of the state that week, and this day looked ominous as well. The young woman was in the water when Thomas stood on the back of the boat and raised his hands toward the sky. 'Here I am,' he shouted. Moments later, a thunderbolt erupted from a nearby cloud and struck his head. The lightning had nowhere to exit, unfortunately, and death was instantaneous. A doctor on a nearby houseboat tried reviving the lawyer, but it was too late.

Ironically, the case with which Thomas was engaged at the time of his death involved an incident on the lake. He was defending Ronald Richie, who had been in a reckless boating accident that killed three people earlier in the year.

A heat-shocked gene

Getting hit by lightning, especially in open spaces, is hardly a new phenomenon. If there is physical opportunity, the lightning will often exit a person's body, leaving large burns in its wake. Getting hit by an electric bolt, or simply receiving some form of thermal energy, is a severe kind of heat shock. To a limited extent, your body has defenses against sudden thermal disturbances, whether they are derived from electrical origins, or other more mechanical sources.

Your body responds to sudden thermal inputs in many ways. There are systemic responses, organ responses, tissue responses and even cellular responses. It can respond to such sudden stress because of certain genetic sequences embedded within your cells. Similar responses occur in every cell that has ever been studied, whether from a jellyfish or an elephant. When a cell is 'shocked,' there is a dramatic change in the expression of certain genes. They encode a family of proteins that provide the stress-induced, or as it is more commonly termed, the heat-shock response.

The change in cells occurs for a specific purpose: to ensure cellular survival under stressed conditions – conditions that if left alone would result in great damage and ultimately cell death. Proteins involved in the heat-shock response are like the National Guard of the molecular world (Figure 43). They have essential roles to play in the manufacture and delivery of certain biochemicals. They even help fold them properly, like a caring Emergency Medical Technician at the scene of a disaster. Their helping role in delivering molecular goods to various cellular compartments is quite ubiquitous.

One of the most interesting research problems in studying the heat-shock response has to do with specificity. Heat-shock genes are turned on selectively, depending upon the type of input received and the tissue undergoing the stress. For example, there is a heat-shock gene that responds to the presence of heavy metals. There is one that responds to psychological stress – and, curiously, only in certain tissues. What do I mean by that? When an animal is caged, stress is induced and heat-shock genes are turned on, but only in selective areas of the adrenal gland and in blood vessels. They are not turned on anywhere else. This property

THE HEAT SHOCK RESPONSE

It's the reaction of the cell to **STRESS**

A cell often undergoes a variety of stressful insults from the environment. These include sudden elevations in temperature, the introduction of heavy metals, pollutants, internally produced metabolic waste products and radiation. Even psychological stress can influence cellular biochemistry.

Cells have found ways to cope with external insults. A dramatic change in the expression of genes can accompany a sudden environmental stress; because it was first observed by altering thermal conditions, the change is known as the heat shock response. New gene products are made, depending upon the kind of stress experienced.

As we age, the heat shock response is greatly diminished. The elevation of helpful gene products is not nearly as dramatic in older organisms when compared to younger ones. This reduction may explain why we lose our ability to respond to acute environmental stresses in later adulthood.

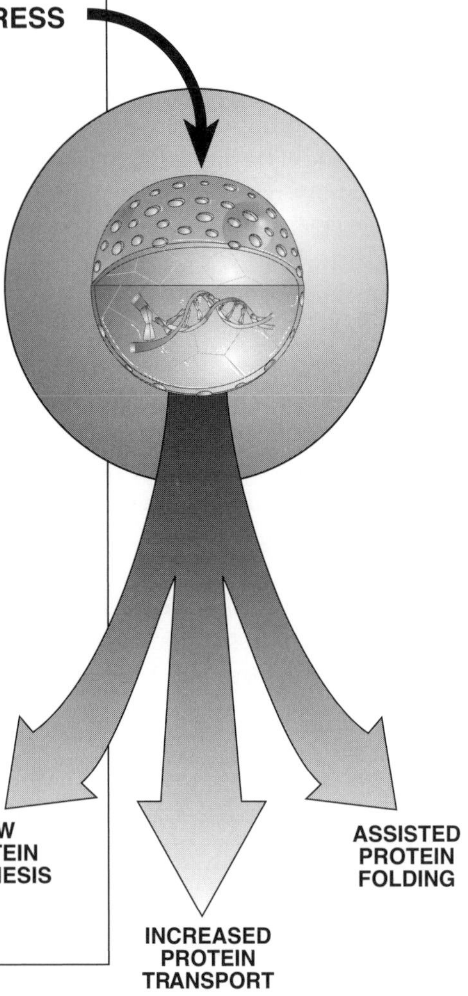

NEW PROTEIN SYNTHESIS

INCREASED PROTEIN TRANSPORT

ASSISTED PROTEIN FOLDING

FIGURE 43

of specificity is involved in a mystery. There are heat-shock genes that don't respond to *any* known cues. We just know they are there. This raises the interesting possibility that there may be cellular stresses our tissues undergo that we have not yet discovered.

The heat-shock genes and aging

The relevance of heat-shock genes to aging has to do with a principle any good stock-room person at an emergency room knows about intimately: proper response depends upon the level of inventory of critical supplies and medicines. It is not different in a cell. The heat-shock response is all about emergency situations due to environmental change. One of the great environmental changes a cell enounters is the aging of the creature in which the cell resides. When a cell suffers damage, could an inadequate supply of heat-shock proteins result in the deterioration normally seen in individuals as they get older? Are heat-shock proteins directly involved in the aging process?

The answer turns out to be yes, although the data for now are mostly correlative. Much of these results have come because of work on a heat-shock protein called HSP70. It has been examined in aging animals and in aging cultures of cells. In both sets of studies, there is a dramatic reduction in the amount of HSP70 as the creature's cells get 'older.' In early adulthood, a shocked organism creates a large amount of this protein. Apply that same shock to an older organism and only a reduced amount of this protein is observed. This means there is an inability of older organisms to respond to acute stresses in their lives. Or at least in their cells. A key defense mechanism against stresses in cells and tissues is actually reduced with age.

Does that mean anything to our discussion of the Error Accumulation theory of aging? It might. One of the great traumas a cell experiences is oxidative stress. This is a fancy term scientists use to describe something we just discussed, the affects of free radicals. Those nasty ROS molecules start destroying critical molecules and structures within cells. Heat-shock genes normally respond to this stress, sending out their National Guards and restoring order to the interior.

But what if The Clock of Ages says the heat-shock gene is to be turned off? That the gene is to be rendered incapable of being copied into an mRNA? Or the mRNA is destroyed before it can make a protein? Then heat-shock proteins and the help they provide would no longer be available to the cell.

In the case of HSP70, that's exactly what The Clock of Ages seems to be saying. If the gene is turned off, the molecular help HSP70 provides is no longer available. The result is that the cell just takes the damage on its molecular chin. And if HSP70 is turned off in lots of cells, the debilitation will be widespread. The cells will eventually lose their function. The altering of the expression of the heat-shock genes may be one factor in the molecular mechanisms of aging.

To be fair, we could never blame a lack of response to oxidative stress solely at the feet of heat-shock proteins. Other genes, whose products are fully capable of repairing damaged tissue, are also turned on when the free radical level gets too high. These proteins have names like glutathione peroxidase, catalase, heme oxygenase, and so on.

But the heat-shock proteins illustrate an important point about later adulthood: *genetic responses to enviromental damage change with age.* There just aren't as many Emergency Response Teams to react to the normal wear and tear our environment places upon us. And the level of help changes with age because of alterations in gene expression. This results in cumulative disrepair. Which results in aging.

Error accumulation

The next set of molecules to be discussed have absolutely nothing to do with lightning-struck lawyers and emergency room responses to trauma. Our next subject concerns something we normally think of sweetly; it has to do with sugar. Only we will not be talking about the taste that stimulates the nerves on our tongue, but rather certain molecules that float inside all of our cells.

Our bodies normally add sugar molecules to certain biochemicals. There are proteins which supervise the addition of glucose (a raw subunit of sugar) to other kinds of proteins, for example. These sugars are added to very specific regions within the protein, which is then said to be 'glycosylated.' Not surprisingly, this process is called glycosylation.

This process is normally highly regulated. Specific molecules are supposed to get specific sugars. If some molecules that were never supposed to get sugar suddenly became 'sweetened,' it could cripple their function. If some molecules that were supposed to get it lost the sweet stuff, their functions could be destroyed as well. No one really knows why, but this process of adding sugars begins to break down in later adulthood. In advancing years sugars are added to molecules that should never get them. Or the ones that normally get them begin reacting strangely. In later adulthood, sugars bind to proteins and to DNA molecules in such a fashion that very complex, unwanted structures are created. These structures are called 'advanced glycosylation end' products or simply, AGE products. There is a great accumulation of AGE products in older cells.

What does this mean to the life of the tissues which amass such compounds? No one really knows. There is a suggestion that the increase in AGE products is associated with certain age-related degenerative diseases. They are responsible for the development of certain complications in diabetes, for example. If AGE products can be removed (this was actually done in diabetic rodents), the animals have greatly reduced secondary complications.

There is also evidence that these products cross-link long-lived extracellular proteins. What does that mean? Recall in our discussion of aging that there are great problems in maintaining the integrity and elasticity of certain tissues. We have laid much of that problem at the feet of molecules like elastin and collagen. If these molecules cross-link too extensively, much of the skin's resiliency is lost, for example. Could AGE products be one of the reasons such changes occur? It is entirely possible that the accumulation of sugar-laden molecules contributes greatly to these protein changes in human tissues.

The accumulation of odd, potentially hazardous biochemicals to normal cells is a hallmark of the Error Accumulation theory of aging. AGE products can be removed by other Emergency Response Teams in the cell. Indeed, elements of the immune system can actually bind to AGE products and send out alert signals. This makes the mystery as to why they are allowed to accumulate with age all the more intriguing. It lends credence to the idea that part of the aging process may be explained by a genetic breakdown. The cellular safety institutions that normally protect a cell can no longer keep up with the damage. As a result, the cell dies. And our tissues get older and older. Eventually, they become too old to function.

The process of communication

There can be no more unfortunate error than a mistake in communication. In human activities, it can change the course of our most serious conflicts or make us laugh the loudest. If there are unfortunate usages or misplaced letters, an entire meaning can change. For example, there was an announcement misprinted from a Lutheran church bulletin in which a key letter – the letter 'g' – was missing from the last word. The announcement read: 'Later this evening there will be hymns sung at a park across the street from the church. Please bring a blanket and come prepared to sin.'

The power of misplaced molecules, or changed sequences within the genetic code, can have just as profound a change – though not nearly so humorous – in biochemical 'meaning.' Such changes are called 'mutations' and some can be quite deadly.

The language of the cell is the genetic code, of course, and the source dictionary belongs to the chromosomes which carry our DNA. As we have seen, the aging process appears to affect many kinds of molecules. Does it create mistakes in the DNA with equal enthusiasm? And do those errors accumulate with passing years, creating increasingly weakened tissues? The answers to these questions depends upon whom you are asking.

To fully answer the questions, a quick review of a cellular process is in order. When we talk about mistakes in DNA, in many ways we are talking about two areas: repair mechanisms and replication. When a chromosome becomes injured, there are mechanisms that repair the faulty DNA. They usually involve scraping out the bad nucleotides (those beads we talked about earlier) and replacing them with new ones. The repair mechanisms are quite enthusiastic. The scraping can involve getting rid of and then replacing many healthy nucleotides just to repair a single damaged one. To reflect this excess excision, scientists have adopted the name 'excision repair' to describe the process.

Mistakes can also occur when the DNA is being replicated. As you recall from an earlier chapter, cell division is a normal part of the cell's life cycle. In order to make two genetically identical cells from one, the amount of DNA must first be doubled. This is done through a process known as DNA replication. The DNA is replicated and each cell obtains a copy.

The point at which errors accumulate in chromosomes involves these two processes. (1) If the repair mechanisms begin to break down, the normal wear and tear on the DNA molecule will not be corrected. (2) If the replication process begins to break down, the DNA will be imperfectly copied; the cell that gets the altered genetic information may be considerably weakened as a result. These mistakes can be deadly to a cell; it could die or even become cancerous (Figure 44).

HOW DNA AGES

As we get older, genetic repair processes break down. Here's what happens to DNA molecules.

Chromosomes suffer environmental wear and tear as the years go by. Repair mechanisms normally keep the DNA in working order. When the DNA doubles (as a cell gets ready to replicate), these repairs are incorporated into the new strands.

The DNA accumulates errors with age. This may be due to inadequate repair processes, mistakes in replication mechanisms or both. Illustrated here are three common errors associated with aging DNA molecules:

1) Deletions. Large chunks of DNA are removed from the chromosome and not replaced.

2) Translocations. Segments of DNA are excised from their normal location and spliced into other areas.

3) Point mutations. Individual nucleotides are changed. The precise order of nucleotides is important to protein manufacture. Alterations of critical proteins can be devastating to the health of a cell.

FIGURE 44

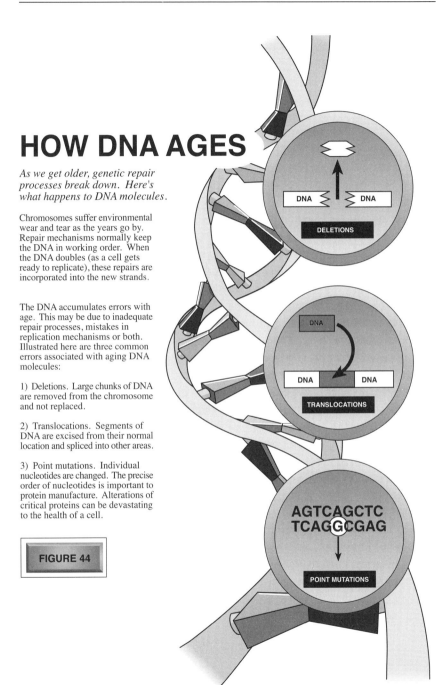

But what about age-related errors?

As mentioned earlier, there is some controversy about an increase in chromosomal aberrations with age. Researchers have looked at young mice and old mice and have found as much as a five-fold increase in chromosomal 'mistakes' with age (gaps and structural rearrangements, for example). Other researchers have examined human cells in the blood and have found an increase in mistakes as well. In one study involving a blood cell type known as a lymphocyte, 60-year-old humans had six times the errors of young adults.

The amount of tiny nicks within the DNA strand, called point mutations, have also been found to increase with age. One study found more than a ten-fold difference in the number of these mistakes when adults over 60 were compared with neonates. Some researchers have followed the structural integrity of very small, specific regions of DNA over the course of time. Mutations collect there like dust particles, and reflect the overall decrease in the accuracy of the genetic code.

Biological research is not often that simple, however. This work has to be contrasted with other studies which have shown very little increases in DNA errors. There are researchers who have demonstrated almost no anomalous DNA rearrangements in higher organisms. Other researchers have looked for and failed to find the significant differences in gaps and mutations. Considering the large differences between individuals, even the point mutation data need some work. There is thus some controversy regarding the nature of DNA integrity with age.

But not too much. The general consensus is that the repair and replication systems break down in later adulthood. Errors collect along the chromosomal strand, and as they are copied into other cells, the errors spread. Because the repair mechanisms (themselves gene products and thus subject to errors) are not as effective, they accumulate as the years go by. Exactly how many errors occur, and how their presence affects our longevity, remains to be seen.

One last gasp of biochemistry

The power of miscommunication can have many consequences in a human cell. We have observed this power in our discussion of the effects of ROS molecules, heat-shock proteins, anomalous sugar additions and unrepaired DNA on senescence. While this might sound like enough to make anyone grow old, it's actually not a complete picture of what is known. Other miscellaneous processes have been described that support, in part, the Error Accumulation theory of aging. One such process has to do with protein degradation and is described below.

The molecular world of the cytoplasm is a cruel place. Many biochemicals, including proteins, are constantly subjected to inspection and review. If a protein is damaged or has outlived its usefulness, it is usually destroyed. This process is called protein degradation and is an efficient means of keeping the cell healthy and unencumbered by useless components.

As we get older, this important housekeeping function begins to fail (Figure 45). Obsolescent or damaged proteins are allowed to hang around. Although the reasons are not clear, it appears that certain molecular 'scissors,' ones that normally chop up the waste, quit functioning. Or are themselves destroyed. Your body also has incinerators that normally receive proteins; there is some evidence that these begin to malfunction with age. This accumulation of useless molecules is observed in cardiac muscle, skeletal muscle, the liver and brain.

So the aging process affects genes and proteins and oxygen and sugars and waste-removal systems. It seems that every tissue and molecule in our body can accrue errors with age. How do we make sense of it all? A summary would help and indeed, several have been published in the research literature. Certain researchers have begun classifying processes into specific categories in an effort to define the Error Accumulation theory more broadly.

Conclusions

What we observe in the Error Accumulation theory is self-killing through neglect. In some way, an important mechanism like repair or glycosylation or waste disposal begins to malfunction. This failure occurs because a gene product, a protein, is no longer able to perform its normal duties. This absence may occur because the gene itself is no longer active.

Which, of course, brings us back to the same old question: what jump-starts the process? In the last chapter, we talked about how genes are switched on and off. We examined the role of promoters, polymerases

The aging of proteins in human cells

As the years go by, the quality and amount of proteins inside cells change. Here's what happens.

Some time ago, it was noticed that the older a cell became, the more protein accumulated in its cytoplasm. How much protein collected depended on the cell type. Cardiac muscle, skeletal muscle, liver and brain tissue all showed dramatic elevations. Since excessive levels can be hazardous to a cell's lifespan, finding out why the accumulation occurred became critically important.

Several biochemical mechanisms normally ensure a steady amount of proteins in younger cells. Described below is what happens when one of those fails.

Proteins in certain nerves accumulate with age

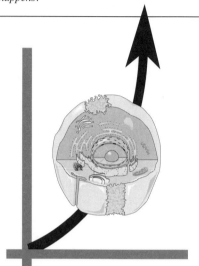

Protein levels dramatically increase in the cytoplasm of certain older cells. There is also a rise in the level of abnormal and dysfunctional proteins.

WHY PROTEINS COLLECT IN AGING HUMAN CELLS

1

Proteins are made in the cytoplasm. They are shuttled to the part of the cell that requires their function.

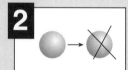

2

When injured or obsolete, the protein is destroyed. This is done by 'incineration' mechanisms within the cell.

3

With age, the incineration mechanisms become dysfunctional. Malformed and useless proteins collect.

FIGURE 45

and transcription factors. In aging research, the next frontier is to understand why these processes fail in critical areas like shock response and replication. The boundaries also begin to blur between the Error Accumulation theory and the hypothesis we discuss in the next chapter, programmed death.

The prescience of playwrights

Oscar Wilde wrote a book about the Error Accumulation theory of aging called the *Picture of Dorian Gray*. He commented on the fleeting nature of youth by likening it to a cigarette: 'A cigarette is the perfect type of a perfect pleasure,' Wilde wrote, 'it is exquisite and it leaves one totally unsatisfied. What more could one want?'

The young man Dorian Gray was given the chance to keep his youth for a long period of time. In keeping with the theory, the errors of biology still accumulated, but not on his body. Rather, they were collected within a portrait of the man. The portrait turned hideous even as Gray turned hedonistic. Eventually, the aging process consumed them both.

Oscar Wilde was not a prescient part-time biologist. The story illustrates nicely, however, the idea of accumulating changes. Relevant to accumulation theory, we find that a great number of molecules are affected in later adulthood. These mistakes come from many sources, pointing once again to the multi-factorial nature of senescence.

The difference, of course, is that the aging process doesn't do it with attitudes. The Clock of Ages does it with biochemicals.

14

Programmed death

If there ever was an example of a deliberately programmed demise of a famous historical figure, it was the events surrounding the death of Napoleon Bonaparte.

Once, the emperor was the most towering figure in all of Europe. His exile to St Helena at the age of 46 was a crushing fall from greatness. In many ways, it illustrates the enormous contradictions of the man, his physical height and his political stature, his recklessness and his careful strategizing, his adversity and his luck. In 1815 (the year of his exile), he was already more myth than man, and his legendary ability to rise from obscurities still produced dread in his enemies. He was banished to the south Atlantic in part from fear. His own words echoed his frustration: 'Ability is of little account without opportunity.'

Arriving on the island, he immediately settled into a life of routine. His health was robust, he had many years to live (so he thought) and he had much of the island at his disposal. Every morning, the former emperor awoke at 9:00, performed his morning toilette, and had breakfast at 10:00. Though he was free to explore, Napoleon never wandered far from his assigned house. Instead, he spent much of his day 'writing' – dictating to his secretary, really – various notes, letters and eventually a will for posterity. He dined at 7:00, read from the books in his library (he was fond of the classics) and retired for the evening at 11:00.

Such a sedentary life for a man who was used to routinely slitting the military jugulars of empires was not beneficial. Soon after he arrived at St Helena, Napoleon's health began to decline.

There has been great controversy about how the military genius really died. A modern analysis of his hair showed the presence of several poisons: arsenic, antimony – even lead. This has caused some historians to speculate that the general's death was 'programmed;' that his food was laced with increasing doses of hazardous substances in a deliberate effort

to kill him. So as not to awaken the cagy man's curiousity, the amount
was increased incrementally. The hope was that eventually a toxic dose
would be administered and Wellington's foe would die a quiet death.

Not everybody agrees with this interpretation of his death, however.
Napoleon probably died a pre-programmed death, but it may not have
come from tainted food. Rather, he may have succumbed to something
in his genes. Two years into his captivity, he was already dying from a
genetic form of cancer.

Napoleon always feared he would expire like his father, who succumbed
to the dread stomach cancer at the age of 38. Several of his relatives had
suffered a similar fate. The first symptoms disrupted Napoleon's routine
in the year 1817. The general started vomiting up what appeared to be
dark granular dirt. It was really the congealed contents of partially
digested blood, tell-tale signs of a malignant stomach ulcer. By 1821, the
year of his death, he was mostly confined to bed.

Doctors on the island tried to treat him. When he was wracked with
stomach and liver pain, the physicians gave him antimony, the standard
treatment of the time, mixed with lemonade. It was a purgative and it
caused Napoleon to fall to the floor in retching agony. The program was
continued, and even augmented with toxic substances such as the mer-
curial laxative called calomel. He was bled, purged and given laxatives
many times, a regimen that did nothing but hasten his death. You can
imagine he was not a happy patient, fighting many times with the doctors
trying to treat him. Such treatment may have been the source of the
poisons found in the general's tissue.

Napoleon fell into unconsciousness on May 5, 1821 and was dead by
5:20 that evening. The last line of his will was 'I die before my time,
killed by the English oligarchy and its hired assassins.'

If the former emperor was talking about his physicians, he was probably
not far off.

This section

Whether one is addressing deliberate exile, insidious poisonings or super-
stitious medical treatment, Napoleon's life illustrates the dramatic power
of programmed curtailment. His strict orders were obeyed even in death –
he was cremated and his ashes returned to his beloved France. We have

samples of his hair, because he also ordered his head shaved prior to cremation. His will dictated that various locks be sent to close friends.

I would like to use the idea of dictation, orders and deliberate death to illustrate the second theory of aging, a process termed Programmed Death (or Programmed Aging). The initial formulation of this idea was in stark contrast to the Error Accumulation theory described in the last chapter. The Programmed Death theory states that aging may occur through a form of pre-planned obsolescence; that certain genetic programs are activated in our cells, which, after a time, cause them to age and die. This is in stark contrast to the Error Accumulation theory, in which death is a lot like one might think of natural land erosion in the face of 'seas of time.'

Proponents have cited examples of deliberate cell destruction in early development as evidence in support of their theory. This includes processes like the death of cells between the digits of humans, the degeneration of muscle cells during insect metamorphosis, the loss of a tadpole's tail in maturity, and so on. We have already discussed the process of apoptosis in human development.

The advent of genetic engineering techniques has brought aspects of this programmed theory into a whole new light. In this section, we are going to explore aspects of the Programmed Aging hypothesis. We will start by describing gene products that govern the cell cycle. Then we will discuss biochemicals whose presence decides the fate of living tissue, and explain how this applies to the aging of whole creatures. Finally we will talk about how the discovery of these genes affects former theories about aging in general, and specifically cell death. In all chapters of this book, I have attempted to illustrate that the aging process is a multi-factorial phenomenon. In this chapter, we will attempt to discover exactly where programmed death fits into this overall scheme.

A genetic basis

Scientists have long suspected a genetic basis for aging processes. Many people have observed that 'long life' tends to run in families. That observation isn't necessarily a scientific one, however. The factors that make human longevity a reality are as multiply environmental as they are hardwired. Nonetheless, other hints have suggested that genes may well be involved in some parts of longevity. And they don't deal with the presence of long life, but with its attenuation.

Science fiction writers and talk-show hosts are very familiar with a syndrome that has been called 'premature aging.' As you know, it is a real phenomenon. People who endure it have an elderly appearance in their youth, because the aging process in these people seems to be greatly accelerated. There are two general types of this premature aging, which are described below.

(1) Hutchinson–Gilford syndrome. This is also called childhood progeria. Children who have this genetic disorder (described as autosomal dominant, which simply means it's not on a sex chromosome and tends to be expressed when present) have the appearance of men and women in their 60s. Chronologically, the subjects have not yet achieved puberty.

(2) Werner's syndrome (or WS). This genetic disease, while harbored in children, does not begin to show itself until after puberty. The first sign that something is wrong comes from the lack of a growth spurt during adolescence. Soon afterward the hair turns gray – and then begins to fall out. Aging spots start to develop. The voice becomes weakened and higher in pitch. Pathologies like osteoporosis and diabetes, usually associated with later adulthood, occur in these patients. So does heart disease. And cancer. The vascular system is so weakened that the usual cause of death of WS patients is a heart attack. The median age of this death is 47 years.

Patients with these genetic diseases do not show all of the symptoms associated with people who age normally. But the fact that such syndromes exist demonstrates that genes can play a dramatic role in the aging process. This can be seen even at the individual cellular level. When tissue is taken from a WS patient and placed into culture, the cells also undergo accelerated senescence.

Another hint that a genetic basis might be behind certain kinds of aging and cell death has to do with apoptosis. As you recall, this word describes the extraordinary and quite deliberate suicide of cells. Their genetic information begins to condense, the cell membrane begins to bleb off and the DNA starts to fragment. The cell dies. You may also remember that this process is vital in embryonic development. The formation of the immune and nervous systems both require deliberate and massive cell death in their construction.

The presence of these phenomena demonstrate that aging may not be as simple as the accumulation of toxic products over a fixed number of years. Some kind of genetic program is operating in these people and in these cells. Understanding aging may be boiled down to discerning the relationship between programmed gene activation, error accumulation and the environment.

If that's the case, then what kinds of programs may be involved? Is there a way we can get at 'aging genes'? Ones whose presence can make cells grow old before their time and whose absence immortalizes them? The startling answer turns out to be yes. Through modern molecular techniques we are beginning to isolate some amazing genetic sequences that appear to mediate our growing old.

Cellular timing

The genes and biochemicals of The Clock of Ages are involved with a cellular timing event that is intimately associated with the cell cycle. To explain the mechanism, we need to discuss a specific structure within the human chromosome, and then talk about the aging of cells in general.

The tip of a human chromosome has biochemical properties so unique that the region has been given a special name. It's called a telomere (Figure 46). Why is it so special? The telomere is involved with the cell cycle, the ability of one cell to divide into two. As you recall, when a cell gets ready to divide, it first has to replicate all of its chromosomes. This it does with great accuracy and precision, until it reaches the tip of the chromosome, that telomere region mentioned above. For complex reasons, the machinery is just too cumbersome to xerox the last little bit. As a result, the tip never gets replicated, and the new cell has fewer genetic sequences at its tip than the old cell.

What does this do to the length of the chromosome? You don't have to be a rocket scientist to know that as the cells continue to divide, the ends of their chromosomes get shorter and shorter. Thus, the more a human cell replicates, the smaller their chromosomes become. As you get older, you have more cells with fewer genetic sequences on their ends. This lack of telomere replication may act like a clock on a time bomb.

A researcher named Olovnikov proposed that this loss of DNA from the tips of chromosomes could be a factor limiting a cell's life span. Why? This self-induced amputation would eventually lead to a loss of essential

The tips of human chromosomes may be

The molecular clock of aging

Called telomeres, the very ends of chromosomes may count-down the life span of a human cell. Here's what telomeres are and how they function in human aging.

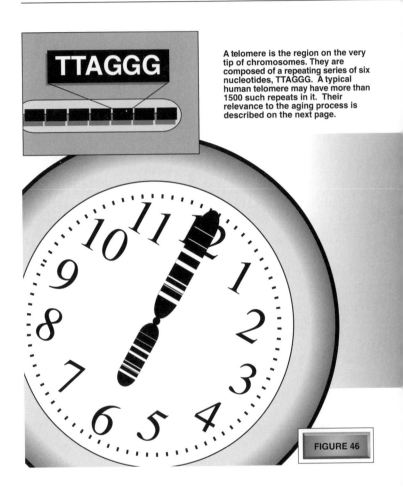

A telomere is the region on the very tip of chromosomes. They are composed of a repeating series of six nucleotides, TTAGGG. A typical human telomere may have more than 1500 such repeats in it. Their relevance to the aging process is described on the next page.

FIGURE 46

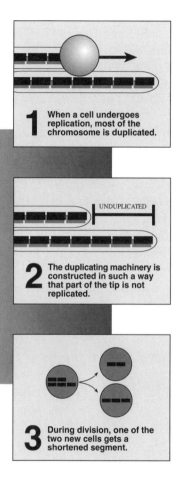

1 When a cell undergoes replication, most of the chromosome is duplicated.

2 The duplicating machinery is constructed in such a way that part of the tip is not replicated.

UNDUPLICATED

3 During division, one of the two new cells gets a shortened segment.

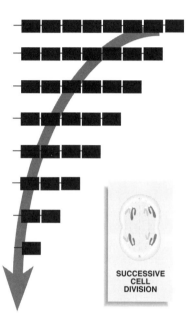

SUCCESSIVE CELL DIVISION

4 As progressive cell divisions occur, the overall chromosomal length gets shorter and shorter. A cell's age can be determined by examining telomere length.

There are studies which suggest this successive loss is involved in aging. When the chromosomes reach a certain length, a self-destruct mechanism is triggered within the cell. This successive loss of telomeres acts like a 'death clock' and strengthens the relationship between aging and programmed cell death.

genes. Indeed, if it were to continue forever, there would theoretically be no chromosomes left. Thus the length of the telomere shows the age of the cell. Like a timed explosion in a bomb, when a certain amount of telomeres are gone, the cell dies.

The idea that a cell may have a molecular clock buried in its genetic information is intuitively very satisfying. But does it have any basis in the reality of aging? The answer appears to be mostly yes. Older cells living within animals do indeed have decreased telomere length. Immortalized tumor cells have found a way to keep their telomeres the same length, regardless of the number of times they divide. There are a number of studies that associate the absence of telomeres with chromosomal instability and cell death.

As ever, more research is required to confirm this timer hypothesis, including reconciling the fact that there are aging organisms that have perfectly intact tips of chromosomes. But cell culture and animal studies suggest, at some level, the involvement of telomeres with aging. They also may be very beneficial in the oncology wards of our hospitals. By targeting the mechanism whereby tumor cells keep their telomeric length intact, new weapons may be created in the fight against cancer.

The presence of a molecular clock at first seems to support the Programmed Death theory of aging. Cells get a certain number of telomeres when they are first 'created' and then, as they replicate, the cells use up their allotment and die. The explicit timing of this mechanism seems to be embedded in the cell's genetic hardware.

The presence of telomeres does not exclude the Error Accumulation theory altogether. Indeed, it may show that they may be two parts of a single process. For example, there is a protein complex that can replicate those chromosomal tips. The enzyme is called telomerase; it's the molecule those tumor cells use to keep their telomeres intact. In most non-tumor cells, the genes that encode the complex are turned off and the cell subsequently ages and dies. Is this turn-off an error? Or the very beginning of a genetic program? The idea that it could be both means that we are losing the boundaries between these two theories – and gaining ever deeper insights into the mystery of human aging.

Genetic programs and stupid thieves

In a moment, we are going to describe another fact of genetic programming. First I'd like to illustrate a principle of the cell cycle by describing not a series of molecules, but some paramilitary car thieves. Ones with double-digit IQs.

In 1990, 12 men from Chula Vista, California decided to steal some cars. Armed with handguns, they piled into a van and drove to The Shipping Company, a business that distributed automobiles all across the United States.

It was a semi-military operation. The thieves arrived at the fence and immediatly subdued and bound the security guards. The criminals then wandered around the lot and picked out the cars they wanted to steal. Nine were taken and driven off the premises. If they had been able to deliver the cars to their 'fence,' this might have been a description of the perfect crime.

That was not to be the case. Unfortunately, the thieves were not aware that each car had been drained of fuel. Only a gallon of petrol remained in each fuel tank. As the fleet of stolen vehicles drove down the freeway, one by one they ran out of gas. One of the cars even suffered a flat tire. This was a bad turn of events, as one of the bound guards soon became free and called the highway patrol. The thieves were all arrested on the highway and the cars were recovered intact.

The aging process

I use this story to illustrate the following principle: the problem with running even ordinary machinery like a car is that so many variables are involved in making it move. There are an equal number of ways it can fail, all resulting in the immobility of the vehicle. For missions that require critical timing, even a silly thing like gas levels makes the difference between success and failure.

The same delicate complexity is true of the missions cells must undertake, even if they are doing something as ordinary as growing and dividing. The reason is that the cells appear to 'run out of gas' – and, as we noticed with telomeres, we may even have a molecular explanation.

But are there other genes involved in the aging of our cells? In aging research, one of the great mysteries is trying to understand why cells stop replicating and start dying. Are there many ways to make a cell stop growing, just like there are many ways to stop the motion of a car? In this section we will explore some of these ideas of cell-cycle regulation. As ever, cancer research, which has studied why cells over-proliferate, has also shed insights into why cells die.

The first gene – *fos*

There are many factors that can start a cell replicating. Because of this fact, there are also many factors that halt the process. To extend our analogy, it is like trying to start a car. The easy and obvious step is to insert the key into the ignition. But all the complex parts necessary to complete the task must be available or the car won't start. The machinery that connects the ignition switch to the electrical system have to be working; pumps and gaskets and rods and tubes must be available to get the gas into the engine, etc. The list is endless, and contrasts with the simplicity of simply turning the key and driving.

In the same way, it can be relatively easy to tell some cells to get growing. If the cell is sitting in a dish, one need only throw in some growth factors and mitosis begins. But the other parts underneath the initial input have to be working as well. A complex machinery of molecules exists, and they must be the proper kinds and in the proper amounts in order to send the signal from the outside of the cell to the nucleus. The nucleus has to have certain parts in working order so that it can incite genes to start this reproductive process, which slowly grinds the reproductive machinery of the cell. It is much more involved than starting a car, and a lot more complex than simply adding a growth factor.

If all the parts are as critical to the growth of the cell as it sounds, you might suspect that the absence of any one of these might cause the cell to shut down. Sort of like having a beautiful car but a dysfunctional ignition system. You might conclude that a valid research direction in aging might be to find out the functional differences between ignition systems of healthy young cells and dying old ones.

And you are exactly right. There are gene products necessary to start a cell in a growth phase which are missing in old cells. One of these is a gene first isolated from cancer cells. It is called *fos* (by convention, whenever a gene sequence is mentioned, the letters are italicized and the name is in lower case. When the protein is mentioned, there are no italics and the first letter is in capital. The gene is *fos* and the protein is Fos, for example). When young cells are given growth factors, the Fos protein is made. It is a transcription factor, which means it can bind to other nuclear proteins and go around waking up other genes involved in cell growth.

Aging cells do not make the Fos protein. As a result, there is no wake-up call to other critical genes in the nucleus. If these other genes are not called up, there will be no cell replication. Even though all the other components may be active and ready to go, the cell nonetheless stops growing. It's just like trying to run a car without any gas. Everything else may be ready to perform their functions, but without the critical one, nothing happens.

There are other genes that seem to act in a similar manner as the *fos* gene. Some of these have been divided into categories, and their effects are so strong that when they are introduced into young growing tissues, the cells stop dividing and die. In many ways, the senescence phenomenon is dominant. That should give pause for some reflection.

Coal-driven people

The Fos protein is only one of a complex mixture of chemicals that can control the cell cycle. A number of genes have been isolated which do much more than just arrest growth in cells. Their affects are so powerful that they can extend or attenuate the life spans of whole organisms.

The cells have many control options, and in order to discuss them, I would like first to describe another accident involving complex machines. This event did not involve stolen cars, but instead confusing dumptruck controls.

In 1984, a man in Pittsburgh, Pennsylvania parked his car on a steep downhill slope. It was directly behind a dumptruck full of coal. While he was gone, another car snuck up behind the first. Seeking to take advantage of the space, the driver of the second car parked so closely that the first car was essentially wedged between his car and the dumptruck. That did not seem to be of concern; the second driver put on his brake and went about his business.

The man from the first car returned and immediately noticed his predicament. How was he to get his car out? Possessing an ability to notice details, the man saw that the dumptruck's door was unlocked. He thought that if he just released the truck's emergency brake, allowing it to roll down the hill a few feet, his problem would be solved. He hopped into the cab, noticed the presence of several levers, grabbed one and pulled.

Unfortunately, it was not the emergency brake. He had mistakenly grabbed the release for the dump lever. The back of the truck raised itself. Frantically the man tried to stop it, but it was too late. The truck dumped

five tons of coal onto the rear automobile, which just happened to be his
own car.

Other genes

The man found that the sticks and levers inside a dumptruck are much
more complex than those in his own car. Pulling the wrong ones can have
dire consequences, and even in lawsuit-happy America, the poor man did
not have any recourse but to take the loss.

I would like to use the story of this unfortunate fellow to illustrate a
point about a newly discovered, remarkable series of genes. The proteins
these genes encode are like levers on a dumptruck. Their stimulation can
cause an entire landslide of aging-related activities to occur. So powerful
are these effects that the life spans of entire organisms are altered.

The first protein is called Apo-1. Apo-1 sticks out of the surface of
cells like levers stick out of the floors of dumptrucks. It is known as a
receptor. This means it requires a molecular 'hand' to grab it, just like a
switch that must be thrown, in order to function. The hands that bind
to these molecular levers are called 'ligands.' An astonishing event occurs
as soon as Apo-1 is grabbed by its ligand. The cell that carries it commits
suicide.

You did not read that wrong. As soon as Apo-1 binds to its ligand,
the cell begins to undergo apoptosis. Its chromosomal material condenses,
the membrane starts to form blebs and the cell crinkles up and dies. The
DNA sequence that encodes Apo-1 protein is a death gene. As soon as
the protein is present and then touched, molecular processes are activated
within the cell that cause the cell to commit suicide. This is a programmed
event. Apo-1 can be placed in any cell, even a young cell, and the cell
will be doomed as soon as the ligand 'hand' is supplied.

Other suicide genes have been isolated. One is called the ICE gene,
for example. If this gene is transferred into animal tissues in a dish, a
funeral is soon held for the cells. The gene can be altered in such a way
that its structure is quite crippled, a process we call mutation. When this
crippled gene is placed into cells which it would normally kill, no death
occurs.

Other suicide genes – and their proteins – are being isolated at a furious
pace. There is a protein called BAX. If a cell has enough BAX present
within its cytoplasm, the cell will die. But if it does not have enough, the

cell will survive. Thus the idea of regulating not only the presence of deadly molecules but also the amount of deadly molecules is very real.

The presence of molecular death angels means that cells can be deliberately programmed to die. The discovery that these are hard-wired into our genetic code is an amazing accomplishment and, somehow, disturbing. It means that the life span of the cell isn't just subject to an uncontrolled build-up of toxic waste. It also means there are deliberate levers that control when disaster is to strike and when it is to be avoided. This may have very important consequences on how we view the senescence of cells and organisms.

Though the isolation of genetic sequences that carry death sentences is extraordinary, the *idea* of deliberate death is not really a surprise. We have known about apoptosis for a long time (recall our discussion of programmed death within immune and nerve cells in Chapter 1). The existence of the death angels begs another question, however. If there are 'death' genes that exert suicidal effects on cell life span, do opposing pieces of genetic information also exist? That is, are there also 'life' genes? If a cell is condemned to die, because of the presence of something like Apo-1, are there molecular judiciary officials who can provide stays of execution? Have they been isolated? The answer to all these questions turns out to be yes. To no one's surprise, the data came from cancer research.

The *bcl-2* gene

Investigators interested in tumor research have been isolating death genes with such fury that each month seems to offer a new set of sequences. But as the days and weeks go by, another set of genes has come into prominence, ones that have the opposite effect. In fact, these genes were first isolated on the basis of their ability to rescue cells that otherwise would undergo apoptosis and die.

One of these rescuing genes is called *bcl-2*. It didn't start out as a hero, however. It was first discovered because a mutation in the gene gives it a bad mood and makes it cancer-causing. The most extraordinary characteristic occurs when *bcl-2* is left intact. It can actually rescue cells from apoptosis. If the gene is placed into a cell that normally has a death sentence on it, the cell will not undergo apoptosis. Instead, the cell will proliferate and survive for normal periods of time.

How does the protein encoded by the *bcl-2* gene work? It affects molecular programs in different ways, depending upon the cell type being examined. In one cell type (nerves), the mechanism of its life-saving action has caused proponents of both the Error Accumulation theory and Programmed Death advocates to leap with pure scientific joy. It joins the two ideas together.

You may recall in our earlier discussion that free radicals within cells cause a lot of damage. Such wastes accumulate as a variety of products we called ROS (reactive oxygen species) molecules. These highly reactive waste products can do damage to the insides of cells. Bcl-2 protein works by acting directly on the synthesis of free radicals. It inhibits their formation, causing a net decrease in the overall amount available inside any one cell. No ROS molecules means no accumulated damage. No damage means no aging and no subsequent death. That's why this work helps to fuse Error Accumulation and Programmed Death. It challenges us to see cellular aging in the same way we see organismal aging – as a complex and multifactorial subset of processes.

There are other rescuers besides *bcl-2* genes. And *bcl-2* is not the whole story of the push–pull world of programmed cell life/cell death. It has been shown, for example, that the life-saving Bcl-2 proteins can bind to death-dealing BAX proteins. The binding neutralizes the effects of each, rather like adding an antacid to a sour stomach. The fate of the cell is thus a struggle between amounts. If there are excess BAX bad guys in the cell, they will bind all the Bcl-2 good guys. The left over BAX will then kill the cell. If there is excess Bcl-2, it will bind up all the BAX. If there is enough leftover Bcl-2, the cell will be saved. The decision to die thus depends on the ratio between these two gene products (see Figure 47).

To make matters more complex, Bcl-2 isn't the whole story. A mouse embryo was constructed that destroyed all its *bcl-2* genes. As one might expect, there was a lot of cell death. But not in nerves. And not enough to stop the formation of an otherwise healthy mouse. Many tissues developed quite normally, implying that other rescuers were around that had not yet been isolated. It is an exciting future direction of this research to find out just how many knights in shining armor exist in our bodies.

Taken together, the world of death-dealing and life-saving genes is a complex, ever growing one. I have only mentioned a few in these pages, and since more are being characterized every week, this information is old even as you read it. Other work in various laboratories around the world have studied not only particular gene products, but also how multiple interactions between many genes can cause the cell to live or die. Three of these gene sequences and/or mechanisms are summarized below.

(1) Senstatin proteins. These gene products can reverse the aging process in skin cells, nerves within the brain and in certain blood vessel cells. BioTech companies hope to apply their biochemistries in an effort to inhibit aging in human skin and as a treatment for atherosclerosis.

(2) Mortality 1 (M1) and Mortality 2 (M2) gene programs. A hierarchical collection of gene products, these programs work in tandem with one another. They were discovered in human cells and utilize such cancer-causing genes as the retinoblastoma gene and something called p53. M1 starts the aging process slowly and then incites M2. M2 finishes off the cells.

(3) Longevity Assurance Genes (LAG genes). These genes have been found in yeast and, more recently, in humans. More will be said about these genes below.

(4) Miscellaneous. Genes that trigger aging in humans have been found on human chromosome 1, chromosome 4 and in other places. The genes for progeria and Werner's syndrome are being hunted down as well. The fact that multiple aging genes exist (there are estimates that the final total may run into the thousands) is not surprising. There are many ways to modulate aspects of the senescence process in a complex organism like a human. No one gene seems to be responsible for all processes in all tissues.

This partial, incomplete list demonstrates that we are just on the threshold of discovering the interaction between toxic waste products and programmed cells. Researchers are only beginning to understand how these various gene products affect aging processes in dish-living cells.

When something is as new (and potentially remarkable) as these results are, they must all be taken with a grain of salt. There is a great distance between events in a dish and events in a real live organism. Do these genes have any relevance to the aging of entire creatures? Can we manipulate such proteins in an experimental animal and get similar remarkable results? In the next section, we will discuss the effects of gene sequences on the aging process in whole creatures. That will give us a clearer perspective on what is a riveting field of research.

The genes of life and death

Gene sequences have been isolated that can control the lifespan of individual cells, and even whole organisms. Here's how a few of them work.

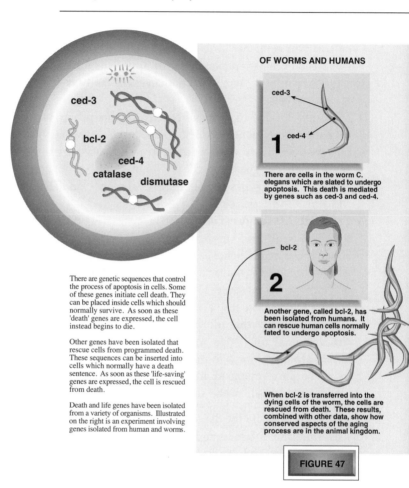

OF WORMS AND HUMANS

1 There are cells in the worm C. elegans which are slated to undergo apoptosis. This death is mediated by genes such as ced-3 and ced-4.

2 Another gene, called bcl-2, has been isolated from humans. It can rescue human cells normally fated to undergo apoptosis.

When bcl-2 is transferred into the dying cells of the worm, the cells are rescued from death. These results, combined with other data, show how conserved aspects of the aging process are in the animal kingdom.

FIGURE 47

There are genetic sequences that control the process of apoptosis in cells. Some of these genes initiate cell death. They can be placed inside cells which should normally survive. As soon as these 'death' genes are expressed, the cell instead begins to die.

Other genes have been isolated that rescue cells from programmed death. These sequences can be inserted into cells which normally have a death sentence. As soon as these 'life-saving' genes are expressed, the cell is rescued from death.

Death and life genes have been isolated from a variety of organisms. Illustrated on the right is an experiment involving genes isolated from human and worms.

COMPETITION BETWEEN GENE PRODUCTS

In certain human cells, Bcl-2 protein can bind to another death-gene product called BAX. The cell can 'read' the internal levels of these two proteins. If there is more free Bcl-2 protein than BAX, the cell will live. If there is more free BAX than Bcl-2, the cell will die.

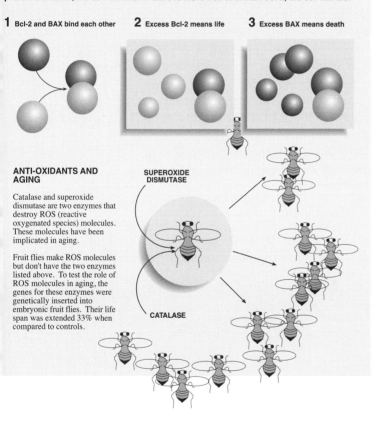

1 Bcl-2 and BAX bind each other   **2** Excess Bcl-2 means life   **3** Excess BAX means death

ANTI-OXIDANTS AND AGING

Catalase and superoxide dismutase are two enzymes that destroy ROS (reactive oxygenated species) molecules. These molecules have been implicated in aging.

Fruit flies make ROS molecules but don't have the two enzymes listed above. To test the role of ROS molecules in aging, the genes for these enzymes were genetically inserted into embryonic fruit flies. Their life span was extended 33% when compared to controls.

SUPEROXIDE DISMUTASE

CATALASE

I will start this discussion by describing a story about reactions to a very human death. This one involved a cruel dictator and his wife.

Death in the Philippines

Imelda Marcos is the wife of former President and now deceased dictator of the Philippines Ferdinand Marcos. She had a hard time adjusting to both losses in her life. When her husband died in Hawaii, she was not allowed to bring the body to the homeland for burial. In the meantime, to prove her love for her beloved Marcos, she changed the shirt on his dead body twice a week. On his birthday in 1990, she gave him a lavish party, with him in attendance. He was propped up in a refrigerated casket. The evening came to a rousing climax when, in front of invited guests, Imelda sang a heartfelt version of Happy Birthday to the corpse.

Eventually, it appeared that she came to grips with her husband's death. The 12 months of 1992 were filled with great natural disasters such as Hurricane Andrew and a variety of Pacific typhoons, volcanic eruptions and storms. Imelda felt that these disturbances were caused by the restless spirit of her dead husband, who could not rest until his remains were buried in the Philippines. She said: 'Look at these typhoons, volcanic eruptions, and now what is happening in America. For the sake of the Republic of the Philippines, for the Filipino people, (we must) put the remains of the president to rest so that these negative vibrations will leave us.'

The subject of aging

Imelda, it seems, had a hard time understanding the finite limitations of the human life span. That they are hereditary, and that some day she will die, may be even further from her mind. These issues of heredity and life span are central not only to an understanding of the human grief response, but also to the research on biochemical aging. In this section, I would like to use the idea of a corpse's perpetuity to link what we know about the genes of aging with their life spans. We will discuss work done not on Pacific ex-dictators, but on fruit flies, yeast and tiny little worms.

A strong genetic component to aging can be observed by simple classical genetics. It illustrates the fact that how long you live can depend on how long your parents lived. These ideas have been given wings with the mating of fruit flies under laboratory conditions. The normal life span of these insects is 40 days. But if you allow only the longest lived of each generation to mate, you can double the life span to 80 days in a remarkably small number of generations. In these experiments, some of the insects live almost six months. If such principles could be applied to humans, it would take about 1500 years to see the results. But the data would be astonishing: our progeny would be living between 400 and 700 years.

As observed in the previous section, this relationship between genes and longevity has taken a much more specific turn. To many researcher's relief, the genes that mediate aging in cultured cells also seem to work in organisms. Those yeast LAG genes we discussed earlier are examples of this association. They can extend the life span of a yeast cell by more than 30%. Such genes have been found in humans as well. Other genetic sequences associated with cancer in human cells also exist in yeast. When the cancer-causing forms of these genes are inserted into yeast, the life spans of the tiny creatures are doubled in length. What is seen in cells can also be observed in whole organisms, albeit very tiny ones. Genetic sequences really can affect The Clock of Ages embedded within creatures.

As the worm turns

A great deal of data have been derived not just from single cells and insects, but also from an extraordinary worm. The size of a comma, this worm has a large name that almost sounds like a scientific compliment – *Caenorhabditis elegans* (mercifully shortened to *C. elegans*, or just 'the roundworm'). This nematode has some extraordinary properties. It defecates every 45 seconds. It tries to mate with everything it sees (it will even try to mate with its own head). Because of the worm's comparatively simple anatomy and biochemistry, its aging process has been amenable to solid scientific inquiry. The results have been surprising.

Like humans, the normal development of the worm involves cell death as well as cell life. Two genes have been found which mediate this death; they are termed *ced-3* and *ced-4*. If these genes are turned off, cell death doesn't occur. As in humans, the worm also has a gene that puts the brakes on this cellular death. This gene is called *ced-9*. If this gene is inactivated, cell death occurs everywhere. If this gene is present, cell death does not occur. When the structure of Ced-9 was determined, the researchers got a giant surprise. It looked just like Bcl-2, the life-saving protein in human cells. *And when they put the human* bcl-2 *gene into doomed worm cells, those cells were rescued just as if they were human cells!*

This is a dramatic finding. The reason is that while there is a great evolutionary distance between worms and humans, they share a similar aging process. Or rather, they share certain functions in similar molecules. The cooperation suggests a dramatic conservation of death processes throughout the animal kingdom. And it strengthens credibility when we attempt to forge research directions in humans by first doing experiments with animals.

Investigations into this roundworm have begun to change the way researchers view the multi-factorial processes of senescence. There is a gene found in the roundworm called *daf-2*. When this one gene is mutated (destroying its function), the life span of *C. elegans* is doubled. The *daf-2* gene can be termed a 'master gene' of aging. This is because its protein can control the activities of many other genes. As we have seen over and over again, aging is the result of many genetic processes. In the round-worm, at least, we have found that many can be under the control of a single master switch.

A human program

In some ways, these switches remind me not of biochemical events but of historical ones, especially the French general mentioned at the top of this section. The actions that led to his demise were actually created by a single decision, somewhat like altering the functions of a single round-worm gene. I will explain below.

In 1812, Napoleon was reeling from a tactical blunder. He had just returned from a disaster in Russia, starting with an invading force of half a million men and returning with only 10 000. With all of Europe begin-ning to band against this erstwhile Emperor of France, the French people took a hint. Napoleon was sacked, his government was toppled, and he was sent to live in friendly exile in Elba, an island off the coast of Italy.

The single decision that led to Napoleon's death came shortly after-ward. Such was the power of the little general that when he left the island, returning to France, he exposed his breast to the French people. 'Strike me down,' he said to their surprised and suddenly nostalgic populace, '. . .or follow me!' They decided, of course, to follow him. Which was a little bit like turning on a death gene.

In 1815, Napoleon led his soldiers – and himself – to the tragic doom of Waterloo. Over 42 000 men lost their lives in the greatest land war Europe had seen to date. It was also the end of the emperor. Napoleon was banished to the south Atlantic, where he had a fatal rendezvous with a cancer gene, six years later.

Conclusions

Like the fatal decision to activate a glory-hungry general, genes exist which can tell cells, and indeed entire organisms, to grow old and die. Not so long ago, the idea that such genes existed would have seemed ridiculous. Human bodies were likened to a cliff on a seashore, constantly being eroded through the mysterious forces of 'time.' The Error Accumulation theory grew out of much of this thinking, cementing into our minds the notion of a fixed, inexorable life span.

The tools of molecular biology began to change this thinking. The idea of Programmed Death, complete with deliberate activation of suicide genes, came into fashion. To be fair, Error Accumulation as the *sole* explanation for organismal aging was always a troubled thought. It had been noted for many decades that cell death in embryonic organisms was the manner by which many vital organs were constructed. But linking processes that were primarily concerned with the beginnings of life to processes primarily concerned with the endings of life still seemed a little far-fetched. It wasn't until the genetic sequences were isolated that such hypotheses came down to earth. It was then easy to incorporate their biochemistries into a solid developmental program theory.

As more molecular data were published, it became obvious that both mechanisms were important. Research into the aging process involved asking questions like 'why does that gene turn off at a certain period of time?' 'What alters the levels of proteins in that cell?' 'How do these changes affect the life span of an organism?' These are the kinds of questions we ask today. The explanation involves both theories and neither theory. Toxic waste products accumulate because genes shut off. Genes shut off because toxic waste products accumulate. Answering the question of which comes first, and what cells employ which hierarchy, are the research directions of the future.

Tucked within the notion that aging has a partial genetic basis is a bombshell of a question, however. If the manipulation of these genes in lower organisms can really result in an alteration of life span *and* these processes are conserved from roundworm to human . . .

. . . will we some day deliberately be able to extend the life span of ourselves and our children?

The answer to that startling question is 'perhaps.' We may need only replace the words 'some day' with the words 'soon.' Or 'now,' depending on who you talk to. We will pick up this idea in our last chapter.

15

Winding back the clock

Alfred Bernhard Nobel has come as close as any human to achieving immortality.

Born in Stockholm, Sweden in 1833, this chemist, who was also the author of the prize that bears his name, grew up to be a man of diverse interests. He was a scientist, inventor and first-rate industrialist, amassing a fortune worth almost $9 million by the time of death. Most of this he threw into an account to fund the prize.

Nobel's occupational interest, however, was explosives and munitions. He made a number of important contributions, including the invention of dynamite. But he was interested in any kind of chemical that could explode. One of these research avenues caused a great tragedy to occur in his life. It may have been part of the reason he established the prize.

Alfred Nobel was researching the properties of nitroglycerin, a volatile explosive of great power. Though he didn't invent the stuff (the honor going to an Italian chemist), Nobel did discover a solution to a very hazardous problem. As you may be aware, nitroglycerin is so explosive that the slightest physical jar can detonate it. Many experimenters were maimed trying to understand the nature of this curious instability. Still more were frustrated because it was a very useful explosive, but could not be transferred from one place to another without great risk. Nitroglycerin remained an oddity of the chemist's research bench for a decade and a half.

That's when Nobel became interested. In 1860, he found that if he carefully dripped into the volatile chemical a gooey concoction of glycerol (a starchy liquid that looks a lot like transparent syrup), the explosive could be safely handled. It was quite an achievement. From it, Nobel was able to raise enough money to build the world's first nitroglycerin factory.

In human terms, it was costly progress. One day the factory blew up, killing five people. Among the dead was Alfred's younger brother Emil.

Nobel decided that his next factory would be built in isolation. He found a barge big enough to hold a chemical manufacturing plant, and moored it in the middle of an isolated lake.

The effect of this tragedy, and the use of his work in warfare, embittered Nobel in later adulthood. It also motivated him. In the year before his death, he wrote an extraordinary will. Regarding the disposition of his cash assets, his will indicated:

> The capital shall be invested . . . the interest shall be annually distrib-
> uted in the form of prizes to those who, during the preceding year,
> shall have conferred the greatest benefit on mankind . . . in awarding
> the prizes no consideration whatever shall be given to nationality of
> the candidate.

The will specified five prizes initially: world peace, medicine, physiology, physics, chemistry and literature. In 1968, the Bank of Sweden added a sixth, economics. It has become research science's highest award, and has achieved for the chemist a characteristic many have sought. The Nobel prize, first issued in 1901, will probably be awarded for many generations to come. It has given Alfred's original fortune a self-sustaining life of its own. It has given him a form of immortality.

An introduction to the section

In this chapter, I would like to use Alfred Nobel's interesting form of immortality to discuss the subject of extended longevity in complex crea-tures. You recall in the last chapter that we examined molecular clocks, death-dealing genes and life-giving proteins. We have seen how these enormously complex molecules work together to construct embryos and kill adults. By altering the functions of some of these genes, we have increased the life span of cells and creatures, or prematurely caused their deaths. The questions that remain to be asked are: can such work be applied to human beings? Can we deliberately change our life span? Do we need fancy chemicals, or can lifestyle changes help us?

The ability to prolong the human life span is the subject of this section. We will begin this discussion of prolongation by examining changes in life style, including the effects of exercise and diet on organismal lon-gevity. Then we will talk about miscellaneous hormonal interventions that may someday dramatically affect the human life span. Finally, we will discuss the role of genes in the aging process, relating what we have seen in simpler organisms to our own longevity.

Before we begin I'd like to comment on the nature of the data presented in this section. Our wish for immortality is rebirthed in every generation that first confronts its will to survive. This strong desire has led to a fantasy-filled jungle of mythology regarding what can and cannot make us live longer. To see this clearly, we must sharpen the machete of the scientific literature and cut through these vines, some of which have been around since ancient times. Even in the 20th century, there have been enough spurious claims about extending human life to fill an encyclopedia. In many ways, the false claims have obscured some truly remarkable progress; modern research has some potentially enormous things to say about human longevity. Made all the more exciting because the data are rooted not within our wills, but within our cells. We will begin by addressing the power of lifestyle.

The role of exercise in extending longevity

By taking a tour of various organ systems and tissues, we will examine how exercise slows down some of the effects of later adulthood. It must be emphasized that increased physical activity is not an explosive that, when detonated inside The Clock of Ages, destroys it. In fact, there are tissues that can be severely damaged if an improper physical pace is instituted. Rather, an exercise regimen is beneficial to human health as we age, dramatically improving the quality of our seniority. As we'll see, however, it is no guarantee of increased longevity.

Skin

The aging process discussed in Part Two affects all three layers of our skin. Cells on our outer surface (epidermis) slough off at a greater rate than they are replaced. Proteins within the middle tissues (dermis) form rigid cross-linkages to each other, resulting in stiffness and loss of flexibility. Even the third fatty layer is affected, losing tissue and redistributing the fat in uneven lumpy layers. When these processes are coupled with a progressive loss of lubricating oils and the ability to sweat, a grim picture emerges. The cumulative effects result in skin that is more susceptible to injury, wrinkles and sagging. It is also much drier.

Can exercise do anything to offset the effects of aging on our skin? The answer is both yes and no. Exercise most directly affects the amount and placement of fat throughout the body, including depots stored under the skin. Without a regular exercise program, aging generally increases the abdominal girth in males by 6–16% and by 25–35% in females. This expansion is due to natural changes in the distribution and accumulation of body fat.

Fortunately, fat-mediated alterations in physical appearance are completely reversible. This fact comes from studies in which older, previously sedentary adults began to incorporate regular physical exercise into their lifestyle. As long as training continued (even into late adulthood), a socially pleasing physique was maintained. Fatty tissue was replaced by lean tissue under the skin, and there was an overall reduction in body fat.

But what about the other aspects of skin aging? Are wrinkles, creases and dryness affected by exercise as well? The answer is more ambiguous. When we discuss the vascular system, we will observe that exercise can improve peripheral circulation. Tissue that remains fed and cleansed of toxic waste doesn't deteriorate as fast. That includes, to a certain extent, our skin.

Unfortunately, there is no exercise that can completely get rid of wrinkles. The Clock of Ages dictates that the turnover rate of skin cells diminishes with age, and the skin naturally weakens. The loss of sebaceous glands and subsequent dryness is inevitable as well. Environmental damage from the 'aging rays of the sun,' not to mention changes in temperature, atmospheric moisture and other environmental pressures, exact their pound of epidermis as well. This kind of deterioration isn't stopped by exercise. It is only halted by reducing dermal exposure to the elements.

In light of this inevitability, commercial preparations have been concocted to artificially reverse some of the effects of aging in skin. Most 'skin care' products that address the aging epidermis contain a chemical called AHA (short for alpha hydroxy acid). This chemical increases the turnover rate of cells in skin, slowing somewhat the normal deterioration. There is probably a statute of limitations on the effectiveness of such products, however. As we have seen from our discussion on telomeres, cells do not have an infinite life span. Instead there is a limit to the number of times they can divide before they either die or incur genetic damage.

Thus AHA is effective as long as there are cells that can respond to its molecular cues and continue to divide.

Overall, exercise can only selectively slow down parts of the aging process on our skin. When there is fat involved, regular physical exercise is the solution to inhibiting otherwise normal changes. But where gene regulation is involved, from the replacement of nerves to the finite capabilities of the cell cycle, the picture is much less clear. In fact, exercise may be irrelevant to these processes. Other activities, such as a reduction in exposure to the environment, or direct chemical intervention, may be required to slow down these effects.

Skeletal muscles

The literature describing the effect of exercise on changing muscle is filled with good news. In fact, of all the tissues in the body, muscles are the most amenable to outside interference.

As you recall from Part Two, there is a normal loss of muscle mass in later adulthood. This may be due in part to a loss of neural responses telling the muscle to move. Such loss can also result from a reduction in blood flow, which changes a muscle's ability to receive oxygen and get rid of waste. These breakdowns result in muscular inactivity. The molecular vultures that normally cannibalize inactive muscle fibers are free to pounce on the proteins in aging muscles.

There is a great deal of evidence suggesting that this deterioration can be slowed. If the muscle remains active, it will not atrophy as quickly. The tissue can maintain or even increase its overall strength, regardless of the age at which it is tested. In fact, aging adults can benefit from exercise, relative to their previous capabilities, at the same rate younger people do.

One study conducted in 69–74-year-old men suggests that gains up to 22% in strength (compared to pre-training achievement) can occur even if the older adult was previously sedentary. This athletic rescue includes the re-establishment of an adequate blood supply. Thus The Clock of Ages shows great mercy to our muscles as the years go by.

But what about the nerves in muscles? Considering a nerve's reluctance to repair itself when damaged, isn't this loss a death sentence on the activity of the muscle it innervates? Unfortunately, the answer is 'probably.' The fact that there is an overall loss in strength and physical endurance even among life-long exercisers points to certain irreversible changes. These alterations may have, in part, neural explanations.

But . . . even with neural interactions, the muscles show great resiliency. We have *lots* of nerves in our bodies. And there is some evidence to suggest that nerve communication in muscle may be augmented even in old age. One study compared gains in the strength of the dominant elbow flexor muscle between younger and older men. Both improved on a regimen of strict exercise. The younger men primarily gained strength through a change in muscle mass. The older men showed no such gain. The study reported their improvement occurred because of a change in 'neural factors.' Exactly what 'neural factors' were not elaborated, unfortunately. It seems likely that a muscle might remain in communication with the brain by using alternative neural pathways – especially as old nerves die off. Until more studies are done, however, these notions must remain in the realm of hypotheses.

Bone

If the muscles can hold up well to certain exercise regimens, how do the structures to which they are attached fare? As you recall, bones age primarily because of a loss of minerals. The osseous tissue not only becomes weaker, it also becomes more brittle. This makes our bones more susceptible to breaking. Because the repair mechanisms are not as active in later adulthood, the breaks are not as easily repaired.

Is there a way to stop this demineralization as we age? An old assumption is that a reduction in activity of the elderly is one cause of reduced bone strength. It follows that an increase in activity will result in an elevation of not only bone mass, but also bone strength.

The literature has been quite mixed in this regard, however. There are studies that suggest an increase in bone mineral content might occur with light to moderate exercise. Other studies say that the increase is nonsense, that bone demineralization is more complicated than can be explained by simple loss of use. The real answer probably lies somewhere in between. The ability to keep minerals stuck to our bones is a complex task, requiring efficient absorption from the intestine and a ready supply of cellular re-modelers. The effect of exercise on these processes has not been greatly studied. In summary, the jury is still out in regards to the effects of exercise on bone strength.

Joints

Where bones and muscles come together is within the great levers that are our joints. If muscles can be greatly improved with exercise, and bones may not be improved at all, how are the joints affected by increased activity? You may remember that joints age because of a complex deterioration of cartilage, tendons and fluid. As collagen and elastin molecules within the ligaments degenerate, the structures begin to fragment. The cartilage begins to change its overall composition, possibly due to the inhibition of certain genetic processes. The fluid in joints which contain liquid begins to thin, as well. As a result, more friction is created. Our joints stiffen and our mobility is decreased.

Is there anything we can do to stop this cellular erosion? The unfortunate answer is no. No amount of exercise can eliminate or reverse the damage that has accumulated over a lifetime of use. The greatest benefit to keeping our joints as flexible as The Clock of Ages allows comes from exercising muscles. If the various muscle groups that support the joints can be strengthened, less stress will be placed upon the weakened tendons, ligaments and joint structures. Since exercising can stimulate blood supply to a local area, the repair processes that still exist may have greater access to stressed tissue. But the verdict from the passage of years is clear. The tendons and ligaments will break down, the cartilage will change molecular composition and the fluid between bones will get too thin. Exercise cannot reverse these processes. Elevated physical activity only serves to increase the likelihood of injury.

Heart, blood and lungs

Some of the most important aging events occur in our heart, blood vessels and lungs. Together these structures form the highways that feed us and take away our molecular trash. Their tissues are extremely resilient, and some can be strengthened even in old age.

If age-related joint damage represents a process we can do the least about externally, changing the senescence of the cardiovascular system represents the opposite. Its potential deterioration is something we can do the *most* about. Extreme interest from the athletic community has spawned nearly two decades of intense research. Admittedly, some of the research has been frivolous. This has been due in part to its popularity in the non-scientific press. But there are enough solid results from controlled experiments that a clear picture has emerged. The results are very encouraging to those approaching their maturity.

Much of this work can be summarized in a single sentence: if a regular program of exercise has been a part of someone's life, the effects of aging in the heart and blood vessels is much less pronounced. Even a training exercise instituted in later life can have enormously beneficial effects. Listed below is a summary of some of the results taken from the scientific literature.

(1) Aerobic capacity. This is the measure of maximum oxygen consumption. Age-related losses can be slowed when a moderate aerobic exercise program has been initiated. Those people who have maintained an active athletic life throughout adulthood have the highest overall capacity within their age-group. Its not fool-proof, however. Even the most physically fit show an age-related reduction in this measurement of cardiovascular efficiency.

(2) Cardiac output. Athletically trained people in later adulthood have a slower resting and submaximal exercise heart rate than sedentary controls. They can also rev up to speed more quickly and maintain a higher rate during exhausting exercise. Perhaps the most significant change, however, has to do with their left ventricles. This chamber of the heart can achieve a higher level of contractility in physically fit individuals than those without the training – as long as those individuals keep exercising. This creates a maximal stroke volume and high cardiac output even during great physical exertion.

(3) Peripheral blood supply. A more efficient energy metabolism in skeletal muscles exists in older people who keep exercising. What does that mean? Simply that these tissues can suck out more oxygen from the blood that coarses through their tissues. More oxygen means a greater efficiency in waste removal. Consequently, there is less damage to tissues by toxic build-up. This is one reason our skeletal muscles stop their atrophy cycle if we decide to exercise.

(4) Changes in blood composition. Lipid metabolism is enhanced in older adults who exercise. This is due primarily to an increase in

that great molecular Noah's ark HDL (high density lipoprotein). These giant conglomerates normally carry lipids from peripheral tissues to the liver for storage. When we exercise, we make more of these transport vehicles. This means there is less free lipid hanging around, which diminishes its opportunity to bind to something else, like the inside wall of a blood vessel. Whereas The Clock of Ages generally demands the presence of more cholesterol, triglycerides and other nasty molecules in the serum, regular exercise can reduce those levels.

(5) Overall benefits to other organs. These changes in circulation efficiency and blood composition can be felt system-wide. Damage to our tissues due to a loss of nutrient supply and inefficient waste removal is greatly reduced when regular exercise is initiated. There are benefits to skin, digestive organs, excretory organs – even reproductive organs – when we stay active later in life.

(6) Changes within the lungs. You might think that exercise would have a great benefit in reducing the effects of aging on the lungs. The assumption is that enhanced movement in and out of the lungs will produce the goal of elevating the oxygenation of the blood. Perhaps the same kinds of benefits exercise gives to the cardiovascular system would be available to the pulmonary (lung) system, right?

Wrong, surprisingly. The literature is actually quite mixed in the effects of exercise on measurements like vital capacity and ventilatory efficiency. The problem is that lung tissue is not made out of muscles which can be worked over. It is made out of delicate skin-like cells which, when damaged, cannot be 'exercised' back to health . The only muscles involved in breathing are those that control inspiration and expiration, and these certainly can be strengthened in an exercise program. This will only increase lung capacity, however. Getting the air into the lungs is a different matter from coaxing oxygen into the blood once the air arrives. Thus there is controversy that exercise does anything to change the age-related deterioration observed in human lung tissue.

Regardless of the lung effects, increased physical activty is a positive thing for older adults. It should encourage us to go out and exercise so that our quality of life in later adulthood can be improved. However, I have to mention a fact that may at first seem like a contradiction. Several decades of research have *not* established an increase in life span from the obvious benefits of exercise. No evidence exists that the maximum

120-years-and-you're-out tenure programming machinery can be altered by anything. People who engage in regular physical activity do not necessarily stay older longer as a result. Let me explain.

The border between the aging process and its relation to exercise can best be seen by asking a few questions. Can an increase in left-ventricle strength halt hair loss in an aging individual? Obviously not. Can increases in peripheral circulation stop menopause from ever occurring? No again. Even an elderly person in the best of shape, who has maintained a life time of vigorous exercise, is not as strong or as aerobically capable as a younger man not nearly so experienced.

What exercise does do for us is to increase the quality of life as we enter old age. Thus, regular physical activity means that we are not as susceptible to certain unpleasant side-effects of deterioration. It also means we are not as susceptible to heart attacks, hypertension and a variety of sedentary-associated diseases. We instead die from different things. Remember, the aging process has components that are as related to gene activation and suppression as they are to physical activity. It does not appear that exercise can ultimately halt the ticking of The Clock of Ages.

The effects of diet

While the data on exercise show a demonstrable effect on the quality of our aging, physical output is not the only activity that can alter the number of years we will live. It is becoming clear that the more calories we consume, the less time we will have on the planet. I use the word 'less' loosely; only in a few cases has this correlation been seen in humans. But the verdict of the laboratory is quite strong. Aging has a great deal to do with diet. *The Clock of Ages seems to be affected by what we eat.*

The effect of diet on aging was first observed in laboratory rats and mice. If the daily food intake is cut by 40 or even 50%, a variety of age-related and life-shortening processes are inhibited. For example, the incidence of diseases of all kinds is reduced. So is the incidence of cancer. The normal age-related decrease in the immune response disappears, and may be responsible for the more vigorous response to disease. Most surprising to the researchers was the effect on longevity. If a laboratory animal was kept on a near-starvation diet, the amount of time it lived compared to normally fed rats increased by almost 50%. This was an extraordinary result, portending mechanisms very different from exercise.

Naturally, the content of the diet was tested. It was determined that these anti-aging effects did not depend on the level of fat in the diet. Nor did they depend on the amount of carbohydrate or even protein. *The variable that mattered to The Clock of Ages was the number of calories in the meal.* The level of vitamins and minerals were held constant to ensure that famine did not reign in the laboratory.

The effects of starvation were not without their side-effects, however. In young animals, there was of course a large decrease in body growth. There was also a delay in the onset of puberty. In adult animals, tissue loss was primarily due to a reduction in overall bodily weight. There was also a decrease in many reproductive capabilities. In old animals, there was a marked reduction in body temperature, a lower blood glucose level and a decrease in whole-body metabolism.

One of the most curious consequences of this diet restriction had to do with changes in hormone secretion. The nerves told specific glands in starved animals to quit secreting biochemicals like growth-hormone, luteinizing-hormone-releasing-hormone and virtually every hormone the pituitary normally secreted into the blood stream. Hormones from target glands (from thyroid, gonads, pancreas, adrenals) were consequently affected.

It has been discovered that this change in hormone levels mediates a good deal of the anti-aging effects. If hormones are externally supplied to starving laboratory animals, they quit living so long. Organ shrinkage is reduced. Depending on the types of hormones re-introduced, reproductive capabilities come back. And, curiously, the incidence of tumors increases. The anti-aging effects are greatly influenced by the presence or absence of hormones in the starving creature.

Does this mean that if we quit eating so much chocolate cake we can live to be 150? While that may be a pleasant reflection, the answer is that no one really knows. There is evidence from diseases which affect glands that normally secrete hormones – especially the thyroid glands – that some effect on longevity occurs. Patients who have decreased thyroid activity (hypothyroidism) tend to live longer than patients with active thyroids. Although these data remain purely associative, they suggest that the hormone notion demonstrated in lab animals may have correlates to humans as well.

But that's just about where the trail grows cold. The mechanisms that convince a starving complex vertebrate animal to hang on for dear life are not well understood. Various processes have been proposed, especially at the cellular level. For example, a reduction in circulating hormones means an altered pattern of gene expression in critical cells throughout the animal. It also means decreased cellular growth and greater survival of telomere sequences (those tips on the chromosomes we discussed in the last chapter). This may somehow result in less wear and tear on body organs and tissues, including less damage to our DNA. But exactly how that translates into more years on earth is a great mystery in aging research.

This idea of input molecules affecting our longevity has been seen elsewhere. One idea that has received a lot of publicity has to do with the Error Accumulation model of aging. The idea has been put forth that longevity may be increased by declaring war on ROS molecules. The weapons? A strong diet of anti-oxidants. Their role in slowing down or even reversing the aging process is discussed next.

The role of anti-oxidants in extending longevity

You recall that the internal accumulation of ROS molecules can damage some very important molecules. ROS molecules are not just the result of genetic processes gone awry in our cells, however. ROS molecules can also be created environmentally, from air pollutants, pesticides, radiation, drugs and other sources. This question has arisen: if ROS molecules can be created externally, are there external supplements in our diet we can use to slow the aging effects mediated by these free radicals? What are the effects of anti-oxidants on human aging?

The answer from the literature is straightforward. Nobody knows. There have been attempts to extend longevity in laboratory animals by adminsitration of such anti-oxidants as vitamin E. No extension was observed. Other anti-oxidants were tried. No extension was observed either. Using a battery of anti-oxidants, researchers have tried to extend the life of other animals, including rats, guinea-pigs and our friend the nematode. Once again the results were negative.

Only when genetic processes were directly manipulated was any effect on longevity observed in complex creatures. I am referring to the fruit fly experiment mentioned earlier. As you recall, foreign ROS scavenging genes were placed into the insect and when it grew up, its longevity was assessed. As a result of the manipulation, the amount of time these creatures lived when compared to controls was doubled. Reducing the number of ROS molecules can thus have a dramatic effect. It may just be that our present route of administering anti-oxidants does not give proper access to the molecules that are killing us.

These experiments demonstrate the value of keeping the Error Accumulation model in our overall assessment of aging. By establishing a framework, several hypotheses have been created that provide a number of research directions for scientists. One interesting finding neurobiologists have uncovered involves ROS molecules and neural death in specific regions of the brain. To describe what I mean, we will discuss a part of the brain called the hippocampus, an area we previously discussed (see Part Two). You may recall that it is involved in memory processing and literally means seahorse.

Cognitive abilities

We have mentioned several times that the selective death of brain cells was an inevitable consequence of the aging process. Loss did not occur in all parts of the brain, and it was not always easy to correlate the change with specific behavioral deficits. We also mentioned that the levels of specific neural transmitters changed within certain regions of the brain. Finally we talked about a phenomenon familiar to everyone over the age of 30. There appears to be a change in our ability to access the memories we have stored in both short- and long-term forms as we get older. Studies have shown that memory loss is one of the greatest fears older adults experience as they confront their maturity.

Can any of these deficits be altered? There is some evidence that the hippocampus may be involved. As we get older, the hippocampus loses a specific neurotransmitter called acetylcholine (ACh). This loss occurs in animals as well as humans. Drugs which can mimic or stimulate ACh production have been given to older rats, monkeys and humans. Their memory capabilities were assessed after treatment.

The results? Most of the drugs did nothing. There were no differences when the memory capabilities of controls were compared to those of the treated subjects. There was one dramatic and very exciting exception. A chemical called physostigmine was given to older animals and humans. Rather than stimulating the production of new ACh, this drug *inhibited* the breakdown of ACh already in the brain. That was the ticket. The memories improved in every organism, including humans, that received the inhibitor. There are some problems with the work, including the potential severity of side-effects. But the fact that a drug with a known biochemistry can in fact alter something as seemingly amorphous as 'memory' creates very exciting research directions for the future.

The role of hormones in extending longevity

As discussed earlier, hormonal loss is an actual fact of the aging process. Even those hormones involved in the maintenance of sexual tissues are affected. These are some of the most powerful forces in the arsenal of The Clock of Ages. The reason is that hormones exert powerful effects over a wide range of tissues. And when these hormones are gone, so are the effects. Discussed next are some data regarding aging and the presence of these hormones, followed by a very interesting experiment.

The most important changes may occur in the hypothalamus, that thumbnail sized structure I called the world's smallest diplomat (see Part 2). The hypothalamus is required to secrete a wide range of extremely critical biochemicals, nearly always in response to input signals. With increasing age, the organ fails to respond to some of these input signals. As a result, it doesn't secrete its critical chemicals. And the body pays for it. Our sex organs quit functioning normally. Our thyroid hormones don't function as effectively and *that* can produce whopping changes in our overall metabolism. Most critically, we quit making growth hormone. We are just beginning to understand how critical circulating levels of this hormone are to the maintenance of bodily functions. In fact, this hormone may hold a great key to human aging.

Hormonal loss also results in decreased immune function. There are hormones manufactured by the hypothalamus that talk directly to the birth place of some of our most important immune weapons. The birth place is called the thymus and the weapons are called T cells. When the thymus doesn't receive its signals from on high, its cadre of immature T cells never grow up to be fighters. Since growth of the thymus is stimulated by growth hormone, its loss is felt by the immune system as well. Other immune cells are also affected.

What does this have to do with longevity? A lot, it turns out. There are drugs which can stimulate to activity even an old, unresponsive hypothalamus in laboratory animals. Consequently, the hypothalamus resumes its normal functions, spitting out its wide-ranging biochemicals, re-exerting a stability to the tissues of the organism. Because of the restored levels, the age-related debilitations do not occur. Sex organs go back to work. The thyroid exerts its global effects and our overall metabolism is restored. The incidence of disease and tumors is reduced as the immune system goes back on line. And most importantly, growth hormone levels are restored. The result is a 50% increase in the longevity of the lab animals that get these stimulants.

Does this mean anything for humans?

The idea that growth hormone may have a great deal to do with human longevity has actually been demonstrated, at least as far as tissue and organ function occur. Described next is a most extraordinary experiment.

The research all had to do with a steroid called hGH, short for human growth hormone. Normally, hGH surges through our bodies, doing a variety of positive things, depending on our age. It helps to build bones. It bolsters the immune system. It aids in wound healing. Our muscles and internal organs maintain their healthy structural integrity because of the presence of hGH.

After the age of 60, this hormone begins to shut down. The process is called hGH menopause, and both genders experience it. It may be one of the reasons why so many bodily functions accelerate their deterioration after that time. This absence suggested an experimental direction to Dr Daniel Rudman at the Medical College of Wisconsin. The many factors that mediate human aging processes might be reversed at least in part, by supplying a hormone which also has multiple effects. What would happen to the aging process if hGH were re-supplied to elderly patients at levels they had as younger adults?

Dr Rudman found a group of 21 elderly men (60s and 70s) willing to try the experiment. He divided them into two groups. The first group was given nothing. They were just asked to come in once a month so that certain age-related biological processes could be assessed (the researchers were looking at how fast they were 'aging'). The other group was given injections of hGH three times a week for six months. This elevated their levels of hGH to the amounts the subjects used to secrete as younger men. It was also given as a precautionary measure. Large doses could produce or aggravate hypertension, enlarge the heart and even affect the joints.

What Dr Rudman found made headlines around the world. The men in the control group exhibited the normal deterioration associated with their age. In fact, some of them lost more muscle, bone and organ tissues than had been expected. But when the group that had been treated with hGH was examined, an extraordinary thing was found. They not only stopped parts of their normal march toward aging, but actually *reversed* certain biological functions. The men regained 10% of their muscle mass. Their skin, normally thinning and becoming more susceptible to abrasion, actually increased in thickness – and by 9%! Internal organs, like the spleen and liver, also gained mass. Several reported soreness in their wrists, but this was because the muscles had expanded to such an extent that the nerves were being stimulated. The overall result? These men had stopped and then reversed a deterioration that accumulated in many years of aging.

To make sure that the data obtained were in fact due to hGH, the injected recipient group was taken off the hormone. Sadly, the youthful characteristics they had acquired during the experiment quickly went away. Other experiments have proved the initial observation, and the idea that life span may be extended in fruit flies, yeast and worms has now also been suggested in humans.

This approach, while fundamentally genetic in nature, is very different from examining a single gene and looking for a particular lengthening or shortening of life span. But it in many ways, it demonstrates the same idea. Aging, like death, is a series of biochemical processes. When these biochemical processes are interrupted, so is the aging process.

Another hormonal trick, more recent in application, has been tried successfully in humans as well. And with the same consequences. The hormone is called DHEA, and is secreted by the adrenal glands. Most humans have DHEA in their bloodstreams by age seven. Its peak levels are seen by age 30 and levels rapidly decline after that. By age 70, DHEA is only at about 10% of its peak.

Dr Etienne-Emile Baulieu at the Krelmin-Bicetre hospital in Paris asked the same question of DHEA that Rudman asked of hGH: could supplementation in older individuals result in reversing some of aging's more pronounced effects? The answer turned out to be yes, at least in his preliminary studies. When small doses were administered daily, there was increased mobility, less joint pain, more restful sleep, a greater sense of psychological well-being, and many other characteristics of aging reversal.

To be fair, the data on growth hormone and DHEA are not without controversy. It might be expected that any 'anti-aging pill' would get a lot of publicity, and considering the complex interactions of hormones on human bodies, a lot of inaccurate reporting as well. The severity of side-effects due to long-term administration of these powerful biochemicals has not been documented, nor have all the positive outcomes been firmly established. Such research only points to a powerful principle: if aging is caused by multi-factorial processes, combatting it may take hormones that are capable of mediating those same processes.

In summary

In this chapter we have sought to identify several of the factors that play a role. We have examined the power of exercise to improve our quality of life, and perhaps have been surprised at how little relevance there is to life span. We have also looked at diet, perhaps being equally surprised to see its great relevance to animal and human longevity. The changes food intake make on us have been exerted primarily at the level of hormones. Our examination of neural processes and chemicals secreted by the brain to the body seem to confirm this role. We have even seen the power of hormonal re-introduction with hGH and DHEA, observing that many aging effects can not only be halted, but also be reversed.

There is great power when human genes turn on and off in response to external and internal cues. This activation and de-activation forms the basis for fusing the ideas of Error Accumulation with a Genetic Program. We see that the aging process can be observed by looking at millions of molecules within a single cell or millions of cells within a single organism. In each we see collections of errors and precisely timed genetic programs. What this means for the future is nothing less than a bombshell. There are active researchers in the field today who think we may soon have protocols that could double or even treble normal human life spans. This we will discuss in the next and final chapter. It would make Alfred Nobel jealous to think we could achieve in our anatomy what he could only achieve in memory. And then someday award a researcher his prize for the effort!

16

Conclusions

The legend is that Joseph of Arimathea, the man to whose tomb the crucified Christ was brought, was actually Jesus' uncle. And a tin miner. He often liked to visit the mines and sometimes he would bring along his more famous Nephew. He remains the patron saint of undertakers and tin miners for these two reasons.

There is an alleged history of Joseph after the crucifixion of Christ, one that plays a mighty role in English folklore. He traveled with Mary Magdalene to the north of France. From there, he set sail to Britain and established a church at Glastonbury. He brought with him a certain chalice from the Last Supper, which was the Holy Grail of Arthurian legend. Since the Bards place King Arthur's own castle at Glastonbury, the birthing of the association – complete with The Quest – has a certain romantic logic to it. Joseph planted his own staff at Glastonbury, which became a hawthorn tree that was supposedly 1650 years old. It was said to miraculously blossom every year on Christmas Eve. This was the kind of Catholic nonsense that so incensed the Puritans. But, just in case, the Puritans chopped it down before they left for America. Purity and immortality figure greatly in these legends. Sir Galahad was the only knight of sufficient moral strength to find the Holy Grail. He became spiritually immortal (and physically unearthly) as a result.

It hearkens to a longing in humankind as old as tears. And the story becomes just one more repository for the wish to beat the final ticking of The Clock of Ages.

The Merlins of the 20th century

New wizards have emerged since the days of Tintagel. But they have traded their suits of armor for lab coats. And they are beginning to ask questions that used to be marinated only in myth. It is the reason so many are interested in longevity issues: can we dramatically extend the life of human beings?

It is easy to get excited about some of the results described in the previous section. The largest hint comes from the fact that for so many years, the mechanisms which keep us 'young' and fit work very nicely. Our cells become damaged due to naturally occurring toxic wastes, just as naturally occurring repair mechanisms come on line. The obvious question is: if the aging process is primarily a repair problem, why not perform on humans the same experiments performed on fruit flies? An equivalent success would result in a human being living over two hundred years.

The problem of course is that aging is not just a repair problem. There are subtle programmable steps in cells that contain both life-giving and death-dealing genes. But even this seems to be amenable to scientific experimentation. We are beginning to understand how cells commit to growth or to suicide. And with this knowledge, we have found two extraordinary properties:

(1) By blocking the death genes, we extend not only the life of the cell where the block occurred, but also the life of the organism that contains the cell.

(2) Genes that control these life-and-death decisions in humans work just as effectively in worms. This means the process is conserved, and gives greater weight to the data we see in animals when we think of our own bodies.

These projections are not lost on the scientific community. There are published scientists who are beginning to sound more like the Chroniclers of an earlier time. Only the new recipes are not snake eyes and powdered bones of saints, but test tubes and fancy centrifuges. Here are three quotes.

The first is from Dr Michael Rose, University of California, Irvine. He's done a lot of work on fruit flies and aging.

> I believe . . . in 25 years we could see the creation of the first products that can postpone human aging significantly. This would be only the beginning of a long process of technological development in which human life span would be aggressively extended. The only practical limit to human life span is the limit of human technology.

The second is from Dr William Regelson, professor of medicine at the Medical College of Virginia.

> With the knowledge that is accumulating now about the nutritional and neuroendocrine aspects of aging, and if we develop ways to repair aging tissues with the help of embryonic cells, we could add 30 healthy years to human life in the next decade. And beyond that, as we learn to control the genes involved in aging, the possibilities of lengthening life appear practically unlimited.

The third is from Dr Michael Jazwinski, Lousiana State University Medical Center. He's done a lot of work on aging in common yeast.

> Possibly in 30 years we will have in hand the major genes that determine longevity, and will be in a position to double, triple, even quadruple our maximum life span of 120 years. It's possible that some people alive now may still be alive 400 years from now.

I must remind you that these are not taken from the pages of English fable. These are established scientists, talking about the future of a field they have dedicated their lives to understanding.

But seriously, folks

Is this opinion about the future of aging research universally held by all scientists? The answer, of course, is no. To understand some of their objections, it is important to remember aspects of the aging process we discussed. At the beginning of this book, I promised to take you on a tour of The Clock of Ages. Let's briefly review some highlights.

Our tour began with an unexpected history. We considered the fact that immortality was not as mythical a biological concept as might have been first assumed. That perhaps what needed changing was not our observations, but our definitions. We observed that defining any death, let alone human death, and the accompanying aging process, can be very difficult.

Even if we could concoct a better-than-working definition, however, we would still be faced with an uneasy undercurrent. We don't really know *why* we age in the first place. The process is obviously untethered from any selective pressure in the natural world. This leaves us with genetic beserkers running amok in the unselected environment of post-reproductive life. And *that* poses the problem of unpredictability, which can make the research exciting, but always hampers our ability to project what we'll find. We are left with aging and dying phenomena that in other animals are optional, that we can't define, let alone do anything else about.

In Part Two, our tour described various components inside the Clock. It became obvious even if we couldn't define what we meant by aging, that we could certainly discover changes experienced in later adulthood. We found that different tissues age at different rates. There are some tissues whose function doesn't seem to be affected by passing years. There are other tissues which are affected so profoundly that their dysfunction is life-threatening. Some tissues have a cell cycle that can be measured in hours, some have a turn-over rate measured in decades. How do we create immortality out of such disparate tissues? Are there multiple ways to create immortality which must somehow cooperate in order to function? Or is the concept of 'forever-life' flawed from the beginning, and what we really mean is 'forever-certain-tissues'?

Our last stop on the tour discussed how these aforementioned components work together to make the Clock tick. We described the presence of toxic waste products, and how their build-up can cause irreparable harm to a living cell. We also discovered the existence of programmed obsolescence. We found that the beginnings of organisms can be filled with death, even as the endings of organisms can be filled with life.

Further examination of this interesting phenomenon allowed us to isolate powerful genetic sequences. We observed death-dealing genes that killed cells and life-saving genes that kept them alive. But we didn't find just one. We found many. And there are researchers existing today who say they number literally in the thousands.

How does one exert the single force of immortality on cells that have within them so many ways to live and die? Many researchers cite the overall success of another biological battle. After declaring a war on cancer decades ago, we still have not found a cure. And the process of aging is infinitely more complicated than the mechanisms underlying cancer. Researchers have gone on record as saying that aging is 'the most complex of all biological problems.' And if we are not close to a cure for the tumors, how much farther away are we from understanding aging?

The stage of research, the detractors say, is a lot like the familiar story of the blindfolded researchers describing the shape of an elephant: there are lots of people feeling various parts of the animal. Some are patting its legs. Some are feeling its tusks. Some are caressing its tail. When they are asked to predict what an elephant looks like, the following responses are given: the leg people say an elephant looks like a cylinder. The tusk people say the elephant is a hard semi-circular creature. The tail people say the elephant looks like a toilet brush. The questioner is then asked to draw a picture of an elephant by assessing the accuracy of the individual investigators.

Are the investigators all correct? No. Are they all incorrect? No, again. The problem is that no one has the complete picture. They're not wrong with their observations, but neither are they right with their projections. The objections of researchers who do not share the optimistic projections can be summarized in a single sentence: we just don't know enough. It may be exciting to look over the vast valley of possibilities, but it may be premature to explain what they are.

The end of The Clock of Ages

Our tour is now done.

There were two main objectives for examining The Clock of Ages. The first was to show how many mechanisms must come into play in order to make an organism age. I have used the word multi-factorial to describe this characteristic about senescence. From head to toe, from proteins to DNA, from birth to death, untold battalions of processes unfold to create the aging of a 60-trillion-celled human.

The tour's other objective was to show our humanity in the face of these processes. That's why our discussion started with the last words I spoke to my mother. Many people have watched people they love grow old and die. Many people get depressed when they observe similar changes within their own body. They sense a loss, somehow, even if they have long understood that worth is so much more than physicality.

Thus it helps to see how others have gone through the process. Whether you are a poet or a general, a writer or a cowboy, you have something in common with everyone who has ever looked in a mirror and made a comparison to their youth. My mother often described this characteristic of the aging process by quoting Amiel:

> To know how to grow old is the master work of wisdom, and one
> of the most difficult chapters in the great art of living.

As a scientist, I often marvel at the *power* of The Clock of Ages, and, even though it's been a long time since her death, how difficult it still remains to write about it all.

Further reading

CHAPTER 1

Gross, J.D. (1994) Developmental decisions in Dictyostelium discoideum. *Microbiological Reviews* **58**(3): 330–51

Kirkwood, T.B. (1992) Comparative life spans of species: why do species have the life spans they do? *American Journal of Clinical Nutrition* **55**(6, Suppl.) 1191s–1195s

Gilbert, S.F. (1985) Colonial eucaryotes: the evolution of differentiation. In *Developmental Biology*, 2nd edition, pp. 23–6, Sinauer Associates, Inc., Massachusetts

Medlen, J. (1993) The black death. *Nursing RSA* 8(11–12): 47–9

Smith, J.M. (1989) Evolution: generating novelty by symbiosis. *Nature* **341**(6240): 284–92

McDougall, J.K. (1994) Immortalization and transformation of human cells by human papillomavirus. *Current Topics in Microbiology and Immunology* **186**: 101–19

Koli, K. and Kaski, O.J. (1992) Cellular senescence. *Annals of Medicine* **24**(5): 313–18

Smith, J.R. and Pereira-Smith, O.M. (1990) Genetic and molecular studies of cellular immortalization. *Advances in Cancer Research* **54**: 63–77

Hickman, F.M. and Hickman, C.P. Sr (1974) *Laboratory Studies in Integrated Zoology*, 4th edition, pp. 107–12, C.V. Mosby Co., St Louis

Young, J.Z. (1991) Light has many meanings for cephalopods. *Visual Neuroscience* **7**(1–2): 1-12

CHAPTER 2

Bright, J., Khar, J. and Khar, A. (1994) Apoptosis: programmed cell death in health and disease. *Bioscience Reports* **14**(2): 67–81

Farber, E. (1994) Programmed cell death: necrosis vs. apoptosis. Modern Pathology 7(5): 605–9

Williams, G.T. (1994) Apoptosis in the immune system. *Journal of Pathology* **173**(1): 11–14

Morris, I. (1992) *Death-ritual and Social Structure in Classical Antiquity*, Cambridge University Press, Cambridge, England

Panati, C. (1989) *Panati's Extraordinary Endings of Practically Everything and Everybody*, Harper & Row, New York

Huntington, R. and Metcalf, P. (1978) *Celebrations of Death: the Anthropology of Mortuary Ritual*, Cambridge University Press, Cambridge, England

CHAPTER 3

Provine, W.B. (1971) *The Origins of Theoretical Population Genetics*, University of Chicago Press, Chicago and London

Medawar, P.B. (1946) Old age and natural death. *Modern Quarterly* **1**: 30–56

Weissmann, A. (1889) *Essays Upon Heredity and Kindred Biological Problems*, pp. 251–333, Clarendon Press, Oxford

Cohen, H.J. (1994) Biology of aging as related to cancer. *Cancer* **74**(4 Suppl.): 2092–100

Sager, R. (1991) Senescence as a mode of tumor suppression. *Environmental Health Perspectives* **93**: 59–62

Hayflick, L. (1987) Origins of longevity. In *Modern Biological Theories of Aging*, ed. H.R. Warner, R.N. Butler, R.L. Sprott and E.L. Schneider, pp. 21–35, Raven Press, New York

CHAPTER 4

Skin

Gilchrest, B.A. (1990) Physiology and pathophysiology of aging skin. In *Physiology, Biochemistry and Molecular Biology of the Skin*, 2nd edition, ed. L.A. Goldsmith, pp. 1425–46, Oxford University Press, New York

Kligman, A.M., Grove, G. and Balin, A. (1985) Aging of human skin. In *Handbook of the Biology of Aging*, ed. C.E. Finch and L. Hayflick, pp. 820–42, Van Nostrand Reinhold, New York

Kenney, R.A. (1982) *Physiology of Aging: A Synopsis*, pp. 31–8, Yearbook Medical, Chicago.

Thorne, N. (1981) The aging of the skin. *Practitioner* **225**: 793–800

Rossman, I. (1977) Anatomic and body composition changes with age. In

Handbook of the Biology of Aging, ed. C.E. Finch and L. Hayflick, pp. 27–45, Van Nostrand Reinhold, New York

CHAPTER 5

Bones

Ralston, S.H. (1994) Analysis of gene expression in human bone biopsies by polymerase chain reaction: evidence for enhanced cytokine expression in postmenopausal osteoporosis. *Journal of Bone and Mineral Research* 9(6): 883–90

Melhus, H., Kindmark, A., Amer, S., Wilen, B. and Lindh, E. (1994) Vitamin D receptor genotypes in osteoporosis. *Lancet* 344(8927): 949–50

Currey, J.D. (1984) Effects of differences in mineralization on the mechanical properties of bone. *Philosophical Transactions of the Royal Society of London (Biology)* 304(1121): 509–18

Horsman, A. and Currey, J.D. (1983) Estimation of mechanical properties of the distal radious from bone mineral content and cortical width. *Clinical Orthopaedics and Related Research* 176: 298–304

Mazess, R.B. (1982) On aging bone loss. *Clinical Orthopaedics and Related Research* 165: 239–52

Avioli, L.V. (1982) Aging, bone and osteoporosis. In *Endocrine Aspects of Aging*, ed. S.G. Korenman, pp. 199–231, Elsevier Biomedical, New York

Joints

Ralphs, J.R. and Benjamin, M. (1994) The joint capsule: structure composition, ageing and disease. *Journal of Anatomy* 184(3): 503–9

Brooks, H. and Fahey, T.D. (1984) *Exercise Physiology: Human Bioenergetics and its Applications*, Wiley, New York

Adrian, M.J. (1981) Flexilbility in the aging adult. In *Exercise and Aging: The Scientific Basis*, ed. E.L. Smith and R.C. Serfass, pp. 45–58, Enslow, Hillside, New Jersey

Munn, K. (1981) Effects of exercise on the range of motion in elderly subjects. In *Exercise and Aging: The Scientific Basis*, ed. E.L. Smith and R.C. Serfass, pp. 167–86, Enslow, Hillside, New Jersey

Brewer, B.J. (1979) Aging of the rotator cuff. *American Journal of Sports Medicine* 7: 102–10

Muscles

Aoyagi, Y. and Shephard, R.J. (1992) Aging and muscle function. *Sports Medicine* **14**(6): 376–96

Kallman, D.A., Plato, C.C. and Tobin, J.D. (1990) The role of muscle loss in the age-related decline of grip strength: cross-sectional and longitudinal perspectives. *Journal of Gerontology* **45**(3): M82–M88

Grimby, G., Danneskiold-Samsoe, B., Hvid, K. and Saltin, B. (1982) Morphology and enzymatic capacity in arm and leg muscles in 78–82 year old men and women. *Acta Physiologica Scandinavica* **115**: 124–34

Tomonaga, M. (1977) Histochemical and ultrastructural changes in senile human skeletal muscle. *Journal of the American Geriatrics Society* **25**: 125–31

Inokuchi, S., Ishikawa, H., Iwamato, S. and Kimura, T. (1975) Age related changes in the histological composition of the rectus abdominis muscle of the adult human. *Human Biology* **47**: 231–49

CHAPTER 6

Brain

Rapp, P.R. and Heindel, W.C. (1994) Memory systems in normal and pathological aging. *Current Opinion in Neurology* **7**(4): 294–8

Rabbitt, P. (1990) Age, IQ and awareness, and recall of errors. *Ergonomics* **33**(10,11): 1291–305

Webb. W.B. (1982) Sleep in older persons: sleep structures of 50 to 60 year-old men and women. *Journal of Gerontology* **37**: 581–6

Botwinick, J. (1977) Intelligence and aging. In *Handbook of the Psychology of Aging*, ed. J.E. Birren and K.W. Schaie, pp. 580–606, Van Nostrand Reinhold, New York

Penfield, W. (1972) The electrode, the brain and the mind. *Zeitschrift fur Neurologie* **201**(4): 297–309

Nerves and neuronal loss

Goldman, J.E., Calingasan, N.V. and Gibson, G.E. (1994) Aging and the brain. *Current Opinion in Neurology* **7**(4): 287–93

Nichols, M.E. (1994) Age-related changes in the neurological examination of healthy sexagenarians, octogenarians, and centenarians. *Journal of Geriatric Psychiatry and Neurology* **7**(1): 1–7

Finch, C.E. (1993) Neuron atrophy during aging: programmed or sporadic? *Trends in Neurosciences* **16**(3): 104–10

Curcio, C.A., Buell, S.J. and Coleman, P.D. (1982) Morphology of the aging central nervous system: not all downhill. In *Advances in Neurogerontology: the Aging Motor System*, ed. J.A. Mortimer, F.J. Prozzola, and G.I. Maletta, pp. 7–36, Praeger, New York

Henderson, G., Tomlinson, B. and Gibson, P.H. (1980) Cell counts in human cerebral cortex in normal adults throughout life using an image analysing computer. *Journal of the Neurological Sciences* **46**: 113–36

CHAPTER 7

Cardiovascular

Hixon, M.E. (1994) Aging and heart failure. *Progress in Cardiovascular Nursing* **9**(1): 4–12

Seals, D.R., Taylor, J.A. and Esler, M.D. (1994) Exercise and aging: autonomic control of the ciruculation. *Medicine and Science in Sports and Exercise* **26**(5): 568–76

Folkow, B. and Svanborg, A. (1993) Physiology of cardiovascular aging. *Physiological Reviews* **73**(4): 725–64

McCardle, W.E., Katch, F.I. and Katch, V.L. (1981) *Exercise Physiology: Energy, Nutrition, and Human Performance*, Lea and Febiger, Philadelphia

Gerstenblith, G. (1980) Noninvasive assessment of cardiovascular function in the elderly. In *Aging: volume 12. The Aging Heart: its Function and Response to Stress*, ed. M.L. Weisfeldt, Raven Press, New York

CHAPTER 8

Lungs

Samaja, M., Rovida, E., Motterlini, R. and Tarantola, M. (1991) The relationship between the blood oxygen transport and the human red cell aging process. *Advances in Experimental Medicine and Biology* **307**: 115–23

Struck, H.J. (1991) Biomechanics and aging. *Zeitschrift für Gerontologie* **24**(3): 121–8

Dempsey, J.R., Johnson, B.D. and Saupe, K.W. (1990) Adaptations and limitations in the pulmonary system during exercise. *Chest* **97**(2 Suppl.): 81s–87s

Brandstetter, R.D. and Kazemi, H. (1983) Aging and the respiratory system. *Medical Clinics of North America* **67**: 419–31

Campbell, E.J. and Lefrak, S.S. (1978) How aging affects the structure and function of the respiratory system. *Geriatrics* **33**(6): 68–78

CHAPTER 9

Digestive system

Iber, R.L., Murphy, P.A. and Connor, E.S. (1994) Age-related changes in the gastrointestinal system. Effects on drug therapy. *Drugs and Aging* **5**(1): 34–48

Evers, B.M., Townsend, C.M. Jr and Thompson, J.C. (1994) Organ physiology of aging. *Surgical Clinics of North America* **74**(1): 23–39

Hosoda, S. *et al.* (1992) Age-related changes in the gastrointestinal tract. *Nutrition Reviews* **50**(12): 374–7

Brauer, P.M., Slavin, J.L. and Marlett, J.A. (1981) Apparent digestibility of neutral detergent fiber in elderly and young adults. *American Journal of Clinical Nutrition* **34**: 1061–70

Goyal, V.K. (1982) Changes with age in the human kidney. *Experimental Gerontology* **17**: 321–31

CHAPTER 10

Senses

Carter, T.L. (1994) Age-related vision changes: a primary care guide. *Geriatrics* **49**(9): 37–42

Stevens, J.C. and Cain, W.S. (1993) Changes in taste and flavor in aging. *Critical Reviews in Food Science and Nutrition* **33**(1): 27–37

Wysocki, C.J. and Pelchat, M.L. (1993) The effects of aging on the human sense of smell and its relationship to food choice. *Critical Reviews in Food Science and Nutrition* **33**(1): 63–82

Steven, J.C. (1992) Aging and spatial acuity of touch. *Journal of Gerontology* **47**(1): 35–40

Gennis, V., Garry, P.J., Haaland, K.Y., Yeo, R.A. and Goodwin, J.S. (1991) Hearing and cognition in the elderly. New findings and a review of the literature. *Archives of Internal Medicine* **151**(11): 2259–64

CHAPTER 11

Reproduction and sex

Schneider, H.D. (1994) Sexuality in old age. *Schweizerische Rundschau für Medizin Praxis* **83**(10): 267–72

Judd, H.L. and Fournet, N. (1994) Changes in ovarioan hormonla function with aging. *Experimental Gerontology* **29**(3–4): 285–98

Roughan, P.A., Kaiser, F.E. and Morley, J.E. (1993) Sexuality and the older woman. *Clinics in Geriatric Medicine* **9**(1): 87–106

Schiavi, R.C. *et al.* (1991) Aging, sleep disorders and male sexual function. *Biological Psychiatry* **30**(1): 15–24

Kaiser, F.E. (1991) Sexuality and impotence in the aging man. *Clinics in Geriatric Medicine* **7**(1): 63–72

CHAPTER 12

Wolffe, A.P. (1994) Transcription: in tune with histones. *Cell* **77**(1): 13–16

Zawel, L. and Reinberg, D. (1993) Initiation of transcription by RNA polymerase II: a multi-step process. In *Progress in Nucleic Acid Research and Molecular Biology*, ed. W.E. Cohen and K. Moldave, pp. 67–108, Academic Press, San Diego

McConkey, D.J. (1993) Cellular signaling in cell death. *New Horizons* **1**(1): 52–9

Dice, J.F. (1993) Cellular and molecular mechanisms of aging. *Phyisological Reviews* **73**(1): 149–59

Thakur, M.K., Oka, T. and Natori, Y. (1993) Gene expression and aging. *Mechanisms of Ageing and Development* **66**: 283–98

CHAPTER 13

Martin, G.R., Danner, D.B. and Holbrook, N.K. (1993) Aging – causes and defenses. *Annual Review of Medicine* **44**:419–29

Orr, W.C. and Sohal, R.S. (1994) Extension of life-span by overexpression of superoxide dismutase and catalase in *Drosophila melanogaster*. *Science* **263**: 1128–30

Lindahl, T. (1993) Instability and decay of the primary structure of DNA. *Nature* **362**: 709–15

Healy, A.M., Mariethoz, E., Pizurki, L. and Polla, B.S. (1992) Heat shock proteins in cellular defense mechanisms and immunity. *Annals of the New York Academy of Sciences* **663**: 319–30

Papaconstantinou, J. (1994) Unifying model of the programmed (intrinsic) and stochastic (extrinsic) theories of aging. The stress response genes, signal transduction–redox pathways and aging. *Annals of the New York Academy of Sciences* **719**: 195–211

Sensi, M., Pricci, R., Andreani, D. and Di Mario, U. (1991) Advanced nonenzymatic glycation endproducts (AGE): their relevance to aging and the pathogenesis of late diabetic complications. *Diabetes Research* **16**(1): 1–9

CHAPTER 14

Sarafian, T.A. and Bredesen, D.E. (1994) Is apoptosis mediated by reactive oxygen species? *Free Radical Research* **21**(1): 1–8

Jackson, M.D. and Evan, G.I. (1994) Apoptosis: breaking the ICE. *Current Biology* **4**(4): 337–40

Stewart, B.W. (1994) Mechanisms of apoptosis: integration of genetic, biochemical and cellular indicators. *Journal of the National Cancer Institute* **86**(17): 1286–96

Korsmeyer, S.J., Shutter, J.R., Veis, D.J. and Merry, D.E. (1993) Bcl-2/Bax: a rheostat that regulates an anti-oxidant pathway and cell death. *Seminars in Cancer Biology* **4**(6): 327–32

Reed. J.C. (1994) Bcl-2 and the regulation of programmed cell death. *Journal of Cell Biology* **124**(1–2): 1–6

Broccoli, D. and Cooke, H. (1993) Aging, healing and the metabolism of telomeres. *American Journal of Human Genetics* **52**(4): 657–60

Blackburn, E.H. (1994) Telomeres: no end in sight. *Cell* **77**(5): 621–3

Martin, G.M. (1989) Genetic modulation of the senescent phenotype in *Homo sapiens*. *Genome* **31**: 390–7

Jazwinski, S.M. (1993) The genetics of aging in the yeast *Saccharomyces cerevisiae*. *Genetica* **91**(1–1): 35–51

CHAPTER 15

Meites, J. (1993) Anti-ageing interventions and their neuroendocrine aspects in mammals. *Journal of Reproductive Fertility* (Suppl.) **46**: 1–9

Buskirk, E.R. (1985) Health maintenance and longevity: exercise. In *Handbook of the Biology of Aging*, 2nd edition, ed. C.E. Finch and E.L. Schneider, Van Nostrand Reinhold, New York

Martin, J.B. and Reichlin, S. (1987) *Clincial Neuroendocrinology*, 2nd edition, volume 28, F.A. Davis, Philadelphia

Rose, M.R. and Nusbaum, T.J. (1994) Prospects for postponing human aging. *FASEB Journal* **8**(12): 925–8

Wood, W.B. and Johnson, T.E. (1994) Aging. Stopping the clock. *Current Biology* **4**(2): 151–3

Rusting, R.L. (1992) Why do we age? *Scientific American* **267**(6): 138–41

Rudman, D. *et al.* (1990) Effects of human growth hormone in men over 60 years old. *New England Journal of Medicine* **323**(1): 1–6

Vu, B.P. (1994) How diet influences the aging process of the rat. *Proceedings of the Society for Experimental Biology and Medicine* **205**(2): 97–105

Birkenhager-Gillesse, E.G., Derkson, J. and Lagaay, A.M. (1994) Dehydroepiandrosterone sulphate (DHEAS) in the oldest, old, aged 85 and over. *Annals of the New York Academy of Sciences* **719**: 543–52

Darrach, B. (1992) The war on aging. *Life* **15**(10): 33–43

Jaroff, L. (1995) New age therapy. *Time* **145**(3): 52

Index